Mastering spaCy

An end-to-end practical guide to implementing NLP
applications using the Python ecosystem

Duygu Altınok

BIRMINGHAM—MUMBAI

Mastering spaCy

Copyright © 2021 Packt Publishing

Group Product Manager: Kunal Parikh
Publishing Product Manager: Ali Abidi
Senior Editor: Roshan Kumar
Content Development Editor: Tazeen Shaikh
Technical Editor: Sonam Pandey
Copy Editor: Safi s Editing
Project Coordinator: Aparna Ravikumar Nair
Proofreader: Safi s Editing
Indexer: Pratik Shirodkar
Production Designer: Joshua Misquitta

First published: July 2021
Production reference: 3211021

Published by Packt Publishing Ltd.
Livery Place
35 Livery Street
Birmingham
B3 2PB, UK.

ISBN 978-1-80056-335-3

www.packt.com

To my mother, Ülker, for her life-long support and endless love. To my sister, for her support and inspiration. To my besties, Umutcan, Simge, and Aydan, for their friendship and support.

Contributors

About the author

Duygu Altınok is a senior **Natural Language Processing (NLP)** engineer with 12 years of experience in almost all areas of NLP, including search engine technology, speech recognition, text analytics, and conversational AI. She has published several publications in the NLP area at conferences such as LREC and CLNLP. She also enjoys working on open source projects and is a contributor to the spaCy library. Duygu earned her undergraduate degree in computer engineering from METU, Ankara, in 2010 and later earned her master's degree in mathematics from Bilkent University, Ankara, in 2012. She is currently a senior engineer at German Autolabs with a focus on conversational AI for voice assistants. Originally from Istanbul, Duygu currently resides in Berlin, Germany, with her cute dog Adele.

About the reviewers

Kevin Lu is currently a student studying software engineering at the University of Waterloo, with experience in full-stack web development, machine learning, computer vision, and natural language processing, and is the founder of the Python package PyATE (Python Automated Term Extraction). His interests include discrete mathematics, data science, algorithmic optimization, and deep learning. In the future, he is interested in pursuing research in NLP with deep learning and applications of it in accelerating academic research.

Usama Yaseen is currently a PhD candidate at Siemens AG (Munich) and the University of Munich. His research interests lie in data-efficient information extraction. Before starting his PhD, he was the lead data scientist at SAP SE, where he led a machine learning team focused on information extraction from semi-structured documents. He holds a master's from the Technical University of Munich in informatics; his master's thesis explored recurrent neural networks with external memory for question-answering systems. Overall, he has worked at Siemens (AG) (on corporate technology research), SAP SE (on machine learning), and Intel Corporation (on software development).

Souvik Roy is an NLP researcher. He primarily works on recurrent neural networks and transformer model compression methodologies such as pruning, quantization, tensor decomposition, and knowledge distillation to reduce the challenges faced by larger models, including longer training and inference times. He is passionate about working with textual data to solve underlying problems. Souvik has a master's in engineering from the University of Waterloo, specializing in text processing. Additionally, he has worked with Scribendi on document summarization and grammatical error correction. Since then, he has been working in diverse industrial research labs.

Carlos Fernando Schiaffin is passionate about analyzing and describing the underlying phenomena of human language. He is an NLP developer currently focused on conversational AI. He has a degree in linguistics and is a self-taught Python programmer. For more than five years, he has been working on NLP systems to try to understand and explain some of the speakers' linguistic behaviors. He started his career as a data tagger and soon went on to design annotation processes for linguistic data in Spanish, English, and Portuguese. Currently, he works with Rasa, spaCy and others, on the development of a conversational AI in Spanish. I thank Duygu Altinok for giving me the chance to participate in this book and my colleagues who always accompany my learning process.

Table of Contents

Section 2: spaCy Features

3
Linguistic Features

4
Rule-Based Matching

5
Working with Word Vectors and Semantic Similarity

6
Putting Everything Together: Semantic Parsing with spaCy

Section 3: Machine Learning with spaCy

7
Customizing spaCy Models

8
Text Classification with spaCy

9
spaCy and Transformers

10
Putting Everything Together: Designing Your Chatbot with spaCy

Other Books You May Enjoy

Index

Preface

spaCy is an industrial-grade, efficient NLP Python library. It offers various pre-trained models and ready-to-use features. *Mastering spaCy* provides you with end-to-end coverage of spaCy features and real-world applications.

You'll begin by installing spaCy and downloading models, before progressing to spaCy's features and prototyping real-world NLP apps. Next, you'll get accustomed to visualizing with spaCy's popular visualizer displaCy. The book also equips you with practical illustrations for pattern matching and helps you advance into the world of semantics with word vectors. Statistical information extraction methods are also explained in detail. Later, you'll cover an interactive business case study that shows you how to combine spaCy features to create a real-world NLP pipeline. You'll implement ML models such as sentiment analysis, intent recognition, and context resolution. The book further focuses on classification with popular frameworks such as TensorFlow's Keras API together with spaCy. You'll cover popular topics, including intent classification and sentiment analysis as well as using them on popular datasets and interpreting the classification results.

By the end of this book, you'll be able to confidently use spaCy, including its linguistic features, word vectors, and classifiers, to create your own NLP apps.

Who this book is for

This book is for data scientists and machine learners who want to excel in NLP as well as NLP developers who want to master spaCy and build applications with it. Language and speech professionals who want to get hands-on with Python and spaCy and software developers who want to quickly prototype applications with spaCy will also find this book helpful. Beginner-level knowledge of the Python programming language is required to get the most out of this book. A beginner-level understanding of linguistic terminology, such as parsing, POS tags, and semantic similarity, will also be useful.

What this book covers

Chapter 1, Getting Started with spaCy, begins your spaCy journey. This chapter gives you an overview of **NLP** with **Python**. In this chapter, you'll install the spaCy library and spaCy language models and explore displaCy, spaCy's visualization tool. Overall, this chapter will get you started with installing and understanding the spaCy library.

Chapter 2, Core Operations with spaCy, teaches you the core operations of spaCy, such as creating a language pipeline, tokenizing the text, and breaking the text into its sentences as well as the `Container` classes. The `Container` classes token, `Doc`, and `Span` are covered in this chapter in detail.

Chapter 3, Linguistic Features, takes a deep dive into spaCy's full power. This chapter explores the linguistic features, including spaCy's most used features, such as **POS-tagger**, **dependency parser**, **named entity recognizer**, and **merging/splitting**.

Chapter 4, Rule-Based Matching, teaches you how to extract information from the text by matching patterns and phrases. You will use morphological features, POS-tags, regex, and other spaCy features to form pattern objects to feed to the spaCy Matcher objects.

Chapter 5, Working with Word Vectors and Semantic Similarity, teaches you about word vectors and associated semantic similarity methods. This chapter includes word vector computations such as distance calculations, analogy calculations, and visualization.

Chapter 6, Putting Everything Together: Semantic Parsing with spaCy, is a fully hands-on chapter. This chapter teaches you how to design a ticket reservation system NLU for **Airline Travel Information System** (**ATIS**), a well-known airplane ticket reservation system dataset, with spaCy.

Chapter 7, Customizing spaCy Models, teaches you how to train, store, and use custom statistical pipeline components. You will learn how to update an existing statistical pipeline component with your own data as well as how to create a statistical pipeline component from scratch with your own data and labels.

Chapter 8, Text Classification with spaCy, teaches you how to do a very basic and popular task of NLP: text classification. This chapter explores text classification with spaCy's `Textcategorizer` component as well as text classification with TensorFlow and Keras.

Chapter 9, spaCy and Transformers, explores the latest hot topic in NLP – transformers – and how to use them with TensorFlow and spaCy. You'll learn how to work with BERT and TensorFlow as well as transformer-based pretrained pipelines of spaCy v3.0.

Chapter 10, Putting Everything Together: Designing Your Chatbot with spaCy, takes you into the world of Conversational AI. You will do **entity extraction**, **intent recognition**, and **context handling** on a real-world restaurant reservation dataset.

To get the most out of this book

First of all, you'll need Python 3 installed and working on your system. Code examples are tested with spaCy v3.0, however, most of the code is compatible with spaCy v2.3 due to backwards compatibility. For the helper libraries such as scikit-learn, pandas, NumPy, and matplotlib, the latest versions available on pip will work. We use TensorFlow, transformers, and helper libraries starting with *Chapter 7, Customizing spaCy Models*, so you can install these libraries by the time you reach Chapter 7.

Software/Hardware covered in the book	OS Requirements
Python >= 3.6	Windows, macOS X, and Linux (any)
spaCy v3.0	Windows, macOS X, Linux (any)
Tensorflow 2.0	Windows, macOS X, Linux (any)
Transformers	Windows, macOS X, Linux (any)
scikit-learn	Windows, macOS X, Linux (any)
pandas	Windows, macOS X, Linux (any)
NumPy	Windows, macOS X, Linux (any)
matplotlib	Windows, macOS X, Linux (any)
Jupyter	Windows, macOS X, Linux (any)

We used Jupyter notebooks from time to time. You can view the notebooks on the book's GitHub page. If you want to work with Jupyter notebooks, that's great; you can install Jupyter via pip. If you don't want to, you can still copy and paste the code into the Python shell and make the code work.

If you are using the digital version of this book, we advise you to type the code yourself or access the code via the GitHub repository (link available in the next section). Doing so will help you avoid any potential errors related to the copying and pasting of code.

Download the example code files

You can download the example code files for this book from GitHub at `https://github.com/PacktPublishing/Mastering-spaCy`. In case there's an update to the code, it will be updated on the existing GitHub repository.

We also have other code bundles from our rich catalog of books and videos available at `https://github.com/PacktPublishing/`. Check them out!

Download the color images

We also provide a PDF file that has color images of the screenshots/diagrams used in this book. You can download it here: `https://static.packt-cdn.com/downloads/9781800563353_ColorImages.pdf`.

Conventions used

There are a number of text conventions used throughout this book.

`Code in text`: Indicates code words in text, database table names, folder names, filenames, file extensions, pathnames, dummy URLs, user input, and Twitter handles. Here is an example: "Finally, the `validation_split` parameter is used to evaluate the experiment."

A block of code is set as follows:

```
import spacy
nlp = spacy.load("en_subwords_wiki_lg")
```

Any command-line input or output is written as follows:

```
wget https://github.com/PacktPublishing/Mastering-spaCy/blob/main/Chapter08/data/Reviews.zip
```

Bold: Indicates a new term, an important word, or words that you see onscreen. For example, words in menus or dialog boxes appear in the text like this. Here is an example: "The following diagram illustrates the distance between **dog** and **cat** and the distance between **dog**, **canine terrier**, and **cat**:"

> Tips or important notes
> Appear like this.

Get in touch

Feedback from our readers is always welcome.

General feedback: If you have questions about any aspect of this book, mention the book title in the subject of your message and email us at customercare@packtpub.com.

Errata: Although we have taken every care to ensure the accuracy of our content, mistakes do happen. If you have found a mistake in this book, we would be grateful if you would report this to us. Please visit www.packtpub.com/support/errata, selecting your book, clicking on the Errata Submission Form link, and entering the details.

Piracy: If you come across any illegal copies of our works in any form on the Internet, we would be grateful if you would provide us with the location address or website name. Please contact us at copyright@packt.com with a link to the material.

If you are interested in becoming an author: If there is a topic that you have expertise in and you are interested in either writing or contributing to a book, please visit authors.packtpub.com.

Reviews

Please leave a review. Once you have read and used this book, why not leave a review on the site that you purchased it from? Potential readers can then see and use your unbiased opinion to make purchase decisions, we at Packt can understand what you think about our products, and our authors can see your feedback on their book. Thank you!

For more information about Packt, please visit packt.com.

Section 1: Getting Started with spaCy

This section will begin with an overview of **natural language processing** (**NLP**) with Python and spaCy. You will learn how the book is organized and how to make the best use of the book. You will then start by installing spaCy and its statistical models and take a quick dive into the spaCy world. Basic operations, general conventions, and visualization are the core attractions of this section.

This section comprises the following chapters:

- *Chapter 1, Getting Started with spaCy*
- *Chapter 2, Core Operations with spaCy*

1
Getting Started with spaCy

In this chapter, we will have a comprehensive introduction to **natural language processing** (**NLP**) application development with Python and **spaCy**. First, we will see how NLP development goes hand in hand with **Python**, along with an overview of what **spaCy** offers as a Python library.

After the warm-up, you will quickly get started with spaCy by downloading the library and loading the models. You will then explore spaCy's popular visualizer **displaCy** by visualizing several features of spaCy.

By the end of this chapter, you will know what you can achieve with spaCy and how to plan your journey with spaCy code. You will be also settled with your development environment, having already installed all the necessary packages for NLP tasks in the upcoming sections.

We're going to cover the following main topics in this chapter:

- Overview of spaCy
- Installing spaCy
- Installing spaCy's statistical models
- Visualization with displaCy

Technical requirements

The chapter code can be found at the book's GitHub repository: `https://github.com/PacktPublishing/Mastering-spaCy/tree/main/Chapter01`

Overview of spaCy

Before getting started with the spaCy code, we will first have an overview of NLP applications in real life, NLP with Python, and NLP with spaCy. In this section, we'll find out the reasons to use Python and spaCy for developing NLP applications. We will first see how Python goes hand-in-hand with text processing, then we'll understand spaCy's place in the Python NLP libraries. Let's start our tour with the close-knit relationship between Python and NLP.

Rise of NLP

Over the past few years, most of the branches of AI created a lot of buzz, including NLP, computer vision, and predictive analytics, among others. But just what is NLP? How can a machine or code solve human language?

NLP is a subfield of AI that analyzes text, speech, and other forms of human-generated language data. Human language is complicated – even a short paragraph contains references to the previous words, pointers to real-world objects, cultural references, and the writer's or speaker's personal experiences. *Figure 1.1* shows such an example sentence, which includes a reference to a relative date (*recently*), phrases that can be resolved only by another person who knows the speaker (regarding the city that the speaker's parents live in) and who has general knowledge about the world (a city is a place where human beings live together):

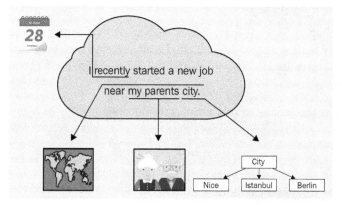

Figure 1.1 – An example of human language, containing many cognitive and cultural aspects

How do we process such a complicated structure then? We have our weapons too; we model natural language with statistical models, and we process linguistic features to turn the text into a well-structured representation. This book provides all the necessary background and tools for you to extract the meaning out of text. By the end of this book, you will possess statistical and linguistic knowledge to process text by using a great tool – the spaCy library.

Though NLP gained popularity recently, processing human language has been present in our lives via many real-world applications, including search engines, translation services, and recommendation engines.

Search engines such as Google Search, Yahoo Search, and Microsoft Bing are an integral part of our daily lives. We look for homework help, cooking recipes, information about celebrities, the latest episodes of our favorite TV series; all sorts of information that we use in our daily lives. There is even a verb in English (also in many other languages), *to google*, meaning *to look up some information on the Google search engine.*

Search engines use advanced NLP techniques including mapping queries into a semantic space, where similar queries are represented by similar vectors. A quick trick is called **autocomplete**, where query suggestions appear on the search bar when we type the first few letters. Autocomplete looks tricky but indeed the algorithm is a combination of a search tree walk and character-level distance calculation. A past query is represented by a sequence of its characters, where each character corresponds to a node in the search tree. The **arcs** between the characters are assigned weights according to the popularity of this past query.

Then, when a new query comes, we compare the current query string to past queries by walking on the tree. A fundamental **Computer Science (CS)** data structure, the tree, is used to represent a list of queries, who would have thought that? *Figure 1.2* shows a walk on the character tree:

Figure 1.2 – An autocomplete example

This is a simplified explanation; the real algorithms blend several techniques usually. If you want to learn more about this subject, you can read the great articles about the data structures: `http://blog.notdot.net/2010/07/Damn-Cool-Algorithms-Levenshtein-Automata` and `http://blog.notdot.net/2007/4/Damn-Cool-Algorithms-Part-1-BK-Trees`.

Continuing with search engines, search engines also know how to transform unstructured data to structured and linked data. When we type `Diana Spencer` into the search bar, this is what comes up:

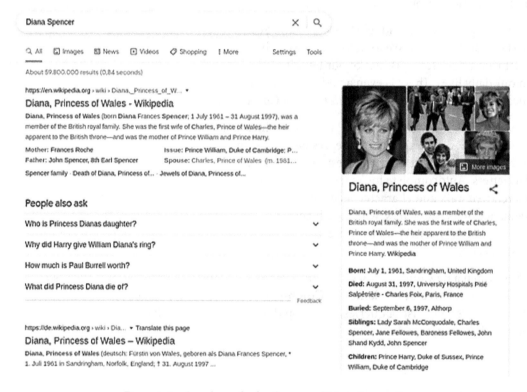

Figure 1.3 – Search results for the query "Diana Spencer"

How did the search engine link `Diana Spencer` to her well-known name *Princess Diana*? This is called **entity linking**. We link entities that mention the same real-world entity. Entity-linking algorithms concern representing semantic relations and knowledge in general. This area of NLP is called the **Semantic Web**. You can learn more about this at `https://www.cambridgesemantics.com/blog/semantic-university/intro-semantic-web/`. I worked as a knowledge engineer at a search engine company at the beginning of my career and really enjoyed it. This is a fascinating subject in NLP.

There is really no limit to what you can develop: search engine algorithms, chatbots, speech recognition applications, and user sentiment recognition applications. NLP problems are challenging yet fascinating. This book's mission is to provide you a toolbox with all the necessary tools. The first step of NLP development is choosing the programming language we will use wisely. In the next section, we will explain why Python is the weapon of choice. Let's move on to the next section to see the string bond of NLP and Python.

NLP with Python

As we remarked before, NLP is a subfield of AI that analyzes text, speech, and other forms of human-generated language data. As an industry professional, my first choice for manipulating text data is Python. In general, there are many benefits to using Python:

- It is easy to read and looks very similar to pseudocode.
- It is easy to produce and test code with.
- It has a high level of abstraction.

Python is a great choice for developing NLP systems because of the following:

- **Simplicity**: Python is easy to learn. You can focus on NLP rather than the programming language details.
- **Efficiency**: It allows for easier development of quick NLP application prototypes.
- **Popularity**: Python is one of the most popular languages. It has huge community support, and installing new libraries with pip is effortless.
- **AI ecosystem presence**: A significant number of open source NLP libraries are available in Python. Many **machine learning** (**ML**) libraries such as PyTorch, TensorFlow, and Apache Spark also provide Python APIs.
- **Text methods**: String and file operations with Python are effortless and straightforward. For example, splitting a sentence at the whitespaces requires only a one-liner, `sentenc.split()`, which can be quite painful in other languages, such as C++, where you have to deal with stream objects for this task.

When we put all the preceding points together, the following image appears – Python intersects with string processing, the AI ecosystem, and ML libraries to provide us the best NLP development experience:

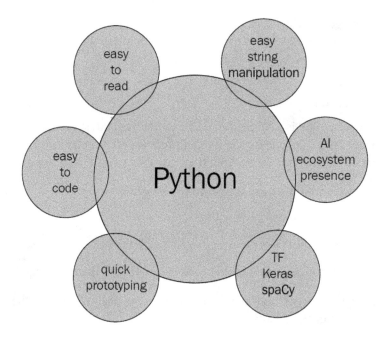

Figure 1.4 – NLP with Python overview

We will use Python 3.5+ throughout this book. Users who do not already have Python installed can follow the instructions at `https://realpython.com/installing-python/`. We recommend downloading and using the latest version of Python 3.

In Python 3.x, the default encoding is **Unicode**, which means that we can use Unicode text without worrying much about the encoding. We won't go into details of encodings here, but you can think of Unicode as an extended set of ASCII, including more characters such as German-alphabet umlauts and the accented characters of the French alphabet. This way we can process German, French, and many more languages other than English.

Reviewing some useful string operations

In Python, the text is represented by **strings**, objects of the `str` class. Strings are immutable sequences of characters. Creating a string object is easy – we enclose the text in quotation marks:

```
word = 'Hello World'
```

Now the `word` variable contains the string `Hello World`. As we mentioned, strings are sequences of characters, so we can ask for the first item of the sequence:

```
print (word [0])
H
```

Always remember to use parentheses with `print`, since we are coding in Python 3.x. We can similarly access other indices, as long as the index doesn't go out of bounds:

```
word [4]
'o'
```

How about string length? We can use the `len` method, just like with `list` and other sequence types:

```
len(word)
11
```

We can also iterate over the characters of a string with sequence methods:

```
for ch in word:
                print(ch)
H
e
l
l
o

W
o
r
l
d
```

> **Pro tip**
>
> Please mind the indentation throughout the book. Indentation in Python is the way we determine the control blocks and function definitions in general, and we will apply this convention in this book.

Now let's go over the more string methods such as counting characters, finding a substring, and changing letter case.

count counts the number of occurrences of a character in the string, so the output is 3 here:

```
word.count('l')
3
```

Often, you need to find the index of a character for a number of substring operations such as cutting and slicing the string:

```
word.index(e)
1
```

Similarly, we can search for substrings in a string with the find method:

```
word.find('World')
6
```

find returns -1 if the substring is not in the string:

```
word.find('Bonjour')
-1
```

Searching for the last occurrence of a substring is also easy:

```
word.rfind('l')
9
```

We can change letter case by the upper and lower methods:

```
word.upper()
'HELLO WORLD'
```

The upper method changes all characters to uppercase. Similarly, the lower method changes all characters to lowercase:

```
word.lower()
'hello world'
```

The `capitalize` method capitalizes the first character of the string:

```
'hello madam'.capitalize()
'Hello madam'
```

The `title` method makes the string title case. Title case literally means *to make a title*, so each word of the string is capitalized:

```
'hello madam'.title()
'Hello Madam'
```

Forming new strings from other strings can be done in several ways. We can concatenate two strings by adding them:

```
'Hello Madam!' + 'Have a nice day.'
'Hello Madam!Have a nice day.'
```

We can also multiply a string with an integer. The output will be the string concatenated to itself by the number of times specified by the integer:

```
'sweet ' * 5
'sweet sweet sweet sweet '
```

`join` is a frequently used method; it takes a list of strings and joins them into one string:

```
' '.join (['hello', 'madam'])
'hello madam'
```

There is a variety of substring methods. Replacing a substring means changing all of its occurrences with another string:

```
'hello madam'.replace('hello', 'good morning')
'good morning madam'
```

Getting a substring by index is called **slicing**. You can slice a string by specifying the start index and end index. If we want only the second word, we can do the following:

```
word = 'Hello Madam Flower'
word [6:11]
'Madam'
```

Getting the first word is similar. Leaving the first index blank means the index starts from zero:

```
word [:5]
'Hello'
```

Leaving the second index blank has a special meaning as well – it means the rest of the string:

```
word [12:]
'Flower'
```

We now know some of the Pythonic NLP operations. Now we can dive into more of spaCy.

Getting a high-level overview of the spaCy library

spaCy is an open source Python library for modern NLP. The creators of spaCy describe their work as **industrial-strength NLP**, and as a contributor I can assure you it is true. spaCy is shipped with pretrained language models and word vectors for 60+ languages.

spaCy is focused on production and shipping code, unlike its more academic predecessors. The most famous and frequently used Python predecessor is **NLTK**. NLTK's main focus was providing students and researchers an idea of language processing. It never put any claims on efficiency, model accuracy, or being an industrial-strength library. spaCy focused on providing production-ready code from the first day. You can expect models to perform on real-world data, the code to be efficient, and the ability to process a huge amount of text data in a reasonable time. The following table is an efficiency comparison from the spaCy documentation (https://spacy.io/usage/facts-figures#speed-comparison):

	ABSOLUTE (MS PER DOC)			RELATIVE (TO SPACY)		
SYSTEM	TOKENIZE	TAG	PARSE	TOKENIZE	TAG	PARSE
spaCy	0.2ms	1ms	19ms	1x	1x	1x
CoreNLP	0.18ms	10ms	49ms	0.9x	10x	2.6x
ZPar	1ms	8ms	850ms	5x	8x	44.7x
NLTK	4ms	443ms	n/a	20x	443ms	n/a

Figure 1.5 – A speed comparison of spaCy and other popular NLP frameworks

The spaCy code is also maintained in a professional way, with issues sorted by labels and new releases covering as many fixes as possible. You can always raise an issue on the spaCy GitHub repo at `https://github.com/explosion/spaCy`, report a bug, or ask for help from the community.

Another predecessor is **CoreNLP** (also known as **StanfordNLP**). CoreNLP is implemented in Java. Though CoreNLP competes in terms of efficiency, Python won by easy prototyping and spaCy is much more professional as a software package. The code is well maintained, issues are tracked on GitHub, and every issue is marked with some labels (such as bug, feature, new project). Also, the installation of the library code and the models is easy. Together with providing backward compatibility, this makes spaCy a professional software project. Here is a detailed comparison from the spaCy documentation at `https://spacy.io/usage/facts-figures#comparison`:

Feature comparison

Here's a quick comparison of the functionalities offered by spaCy, NLTK and CoreNLP.

	SPACY	NLTK	CORENLP
Programming language	Python	Python	Java / Python
Neural network models	✓	✗	✓
Integrated word vectors	✓	✗	✗
Multi-language support	✓	✓	✓
Tokenization	✓	✓	✓
Part-of-speech tagging	✓	✓	✓
Sentence segmentation	✓	✓	✓
Dependency parsing	✓	✗	✓
Entity recognition	✓	✓	✓
Entity linking	✓	✓	✗
Coreference resolution	✗	✗	✓

Figure 1.6 – A feature comparison of spaCy, NLTK, and CoreNLP

Throughout this book, we will be using spaCy's latest release *v3.1* (the version used at the time of writing this book) for all our computational linguistics and ML purposes. The following are the features in the latest release:

- Original data preserving tokenization.
- Statistical sentence segmentation.
- Named entity recognition.
- **Part-of-Speech** (**POS**) tagging.
- Dependency parsing.
- Pretrained word vectors.
- Easy integration with popular deep learning libraries. spaCy's ML library `Thinc` provides thin wrappers around PyTorch, TensorFlow, and MXNet. spaCy also provides wrappers for `HuggingFace` Transformers by `spacy-transformers` library. We'll see more of the `Transformers` in *Chapter 9, spaCy and Transformers*.
- Industrial-level speed.
- A built-in visualizer, displaCy.
- Support for 60+ languages.
- 46 state-of-the-art statistical models for 16 languages.
- Space-efficient string data structures.
- Efficient serialization.
- Easy model packaging and usage.
- Large community support.

We had a quick glance around spaCy as an NLP library and as a software package. We will see what spaCy offers in detail throughout the book.

Tips for the reader

This book is a practical guide. In order to get the most out of the book, I recommend readers replicate the code in their own Python shell. Without following and performing the code, it is not possible to get a proper understanding of NLP concepts and spaCy methods, which is why we have arranged the upcoming chapters in the following way:

- Explanation of the language/ML concept
- Application code with spaCy
- Evaluation of the outcome
- Challenges of the methodology
- Pro tips and tricks to overcome the challenges

Installing spaCy

Let's get started by installing and setting up spaCy. spaCy is compatible with 64-bit Python 2.7 and 3.5+, and can run on Unix/Linux, macOS/OS X, and Windows. **CPython** is a reference implementation of Python in C. If you already have Python running on your system, most probably your CPython modules are fine too – hence you don't need to worry about this detail. The newest spaCy releases are always downloadable via `pip` (`https://pypi.org/`) and `conda` (`https://conda.io/en/latest/`). `pip` and `conda` are two of the most popular distribution packages.

`pip` is the most painless choice as it installs all the dependencies, so let's start with it.

Installing spaCy with pip

You can install spaCy with the following command:

```
$ pip install spacy
```

If you have more than one Python version installed in your system (such as Python 2.8, Python 3.5, Python 3.8, and so on), then select the `pip` associated with Python you want to use. For instance, if you want to use spaCy with Python 3.5, you can do the following:

```
$ pip3.5 install spacy
```

If you already have spaCy installed on your system, you may want to upgrade to the latest version of spaCy. We're using *spaCy 3.1* in this book; you can check which version you have with the following command:

```
$ python -m spacy info
```

This is how a version info output looks like. This has been generated with the help of my Ubuntu machine:

```
============================== Info about spaCy ==============================

spaCy version      3.1.3
Location           /usr/local/lib/python3.8/dist-packages/spacy
Platform           Linux-5.4.0-88-generic-x86_64-with-glibc2.29
Python version     3.8.10
Pipelines          en_core_web_md (3.1.0)
```

Figure 1.7 – An example spaCy version output

Suppose you want to upgrade your spaCy version. You can upgrade your spaCy version to the latest available version with the following command:

```
$ pip install -U spacy
```

Installing spaCy with conda

conda support is provided by the conda community. The command for installing spaCy with conda is as follows:

```
$ conda install -c conda-forge spacy
```

Installing spaCy on macOS/OS X

macOS and OS X already ship with Python. You only need to install a recent version of the Xcode IDE. After installing Xcode, please run the following:

```
$ xcode-select -install
```

This installs the command-line development tools. Then you will be able to follow the preceding pip commands.

Installing spaCy on Windows

If you have a Windows system, you need to install a version of Visual C++ Build Tools or Visual Studio Express that matches your Python distribution. Here are the official distributions and their matching versions, taken from the spaCy installation guide (`https://spacy.io/usage#source-windows`):

DISTRIBUTION	VERSION
Python 2.7	Visual Studio 2008
Python 3.4	Visual Studio 2010
Python 3.5+	Visual Studio 2015

Figure 1.8 – Visual Studio and Python distribution compatibility table

If you didn't encounter any problems so far, then that means spaCy is installed and running on your system. You should be able to import spaCy into your Python shell:

```
import spacy
```

Now you successfully installed spaCy – congrats and welcome to the spaCy universe! If you have installation problems please continue to the next section, otherwise you can move on to language model installation.

Troubleshooting while installing spaCy

There might be cases where you get issues popping up during the installation process. The good news is, we're using a very popular library so most probably other developers have already encountered the same issues. Most of the issues are listed on *Stack Overflow* (`https://stackoverflow.com/questions/tagged/spacy`) and the *spaCy GitHub* Issues section (`https://github.com/explosion/spaCy/issues`) already. However, in this section, we'll go over the most common issues and their solutions.

Some of the most common issues are as follows:

- **The Python distribution is incompatible**: In this case please upgrade your Python version accordingly and then do a fresh installation.

- **The upgrade broke spaCy**: Most probably there are some leftover packages in your installation directories. The best solution is to first remove the spaCy package completely by doing the following:

```
pip uninstall spacy
```

  Then do a fresh installation by following the installation instructions mentioned.

- **You're unable to install spaCy on a Mac**: On a Mac, please make sure that you don't skip the following to make sure you correctly installed the Mac command-line tools and enabled pip:

```
$ xcode-select -install
```

In general, if you have the correct Python dependencies, the installation process will go smoothly.

We're all set up and ready for our first usage of spaCy, so let's go ahead and start using spaCy's language models.

Installing spaCy's statistical models

The spaCy installation doesn't come with the statistical language models needed for the spaCy pipeline tasks. spaCy language models contain knowledge about a specific language collected from a set of resources. Language models let us perform a variety of NLP tasks, including **POS tagging** and **named-entity recognition** (**NER**).

Different languages have different models and are language specific. There are also different models available for the same language. We'll see the differences between those models in detail in the *Pro tip* at the end of this section, but basically the training data is different. The underlying statistical algorithm is the same. Some of the currently supported languages are as follows:

LANGUAGE	CODE	LANGUAGE DATA	MODELS
Chinese	zh	lang/zh </>	3 models
Danish	da	lang/da </>	3 models
Dutch	nl	lang/nl </>	3 models
English	en	lang/en </>	3 models
French	fr	lang/fr </>	3 models
German	de	lang/de </>	3 models
Greek	el	lang/el </>	3 models
Italian	it	lang/it </>	3 models
Japanese	ja	lang/ja </>	3 models
Lithuanian	lt	lang/lt </>	3 models
Multi-language	xx	lang/xx </>	3 models
Norwegian Bokmal	nb	lang/nb </>	3 models
Polish	pl	lang/pl </>	3 models
Portuguese	pt	lang/pt </>	3 models
Romanian	ro	lang/ro </>	3 models
Spanish	es	lang/es </>	3 models

Figure 1.9 – spaCy models overview

The number of supported languages grows rapidly. You can follow the list of supported languages on the **spaCy Models and Languages** page (`https://spacy.io/usage/models#languages`).

Several pretrained models are available for different languages. For English, the following models are available for download: en_core_web_sm, en_core_web_md, and en_core_web_lg. These models use the following naming convention:

- **Language**: Indicates the language code: en for English, de for German, and so on.
- **Type**: Indicates the model capability. For instance, core means a general-purpose model for the vocabulary, syntax, entities, and vectors.
- **Genre**: The type of text the model recognizes. The genre can be web (Wikipedia), news (news, media) Twitter, and so on.
- **Size**: Indicates the model size: lg for large, md for medium, and sm for small.

Here is what a typical language model looks like:

en_core_web_sm

RELEASE DETAILS

Latest: 2.3.1

English mutli-task CNN trained on OntoNotes. Assigns context-specific token vectors, POS tags, dependency parse and named entities.

LANGUAGE	EN English
TYPE	(CORE) Vocabulary, syntax, entities, vectors
GENRE	(WEB) written text (blogs, news, comments)
SIZE	(SM) written text (blogs, news, comments)
PIPELINE ⑦	tagger, parser, ner
VECTORS ⑦	n/a
SOURCES ⑦	OntoNotes 5
AUTHOR	Explosion
LICENSE	MIT

Figure 1.10 – The small-sized spaCy English web model

Large models can require a lot of disk space, for example en_core_web_lg takes up 746 MB, while en_core_web_md needs 48MB and en_core_web_sm takes only 11MB. Medium-sized models work well for many development purposes, so we'll use the English md model throughout the book.

> **Pro tip**
>
> It is a good practice to match model genre to your text type. We recommend picking the genre as close as possible to your text. For example, the vocabulary in the social media genre will be very different from that in the Wikipedia genre. You can pick the web genre if you have social media posts, newspaper articles, financial news – that is, more language from daily life. The Wikipedia genre is suitable for rather formal articles, long documents, and technical documents. In case you are not sure which genre is the most suitable, you can download several models and test some example sentences from your own corpus and see how each model performs.

Now that we're well-informed about how to choose a model, let's download our first model.

Installing language models

Since v1.7.0, spaCy offers a great benefit: installing the models as Python packages. You can install spaCy models just like any other Python module and make them a part of your Python application. They're properly versioned, so they can go into your `requirements.txt` file as a dependency. You can install the models from a download URL or a local director manually, or via `pip`. You can put the model data anywhere on your local filesystem.

You can download a model via spaCy's `download` command. `download` looks for the most compatible model for your spaCy version, and then downloads and installs it. This way you don't need to bother about any potential mismatch between the model and your spaCy version. This is the easiest way to install a model:

```
$ python -m spacy download en_core_web_md
```

The preceding command selects and downloads the most compatible version of this specific model for your local spaCy version.

To download the exact model version, the following is what needs to be done (though you often don't need it):

```
$ python -m spacy download en_core_web_lg-2.0.0 --direct
```

The download command deploys pip behind the scenes. When you make a download, pip installs the package and places it in your site-packages directory just as any other installed Python package.

After the download, we can load the packages via spaCy's load () method.

This is what we did so far:

```
$ pip install spacy
$ python -m spacy download en_core_web_md
 import spacy
 nlp = spacy.load('en_core_web_md')
 doc = nlp('I have a ginger cat.')
```

We can also download models via pip:

1. First, we need the link to the model we want to download.

2. We navigate to the model releases (https://github.com/explosion/spacy-models/releases), find the model, and copy the archive file link.

3. Then, we do a pip install with the model link.

Here is an example command for downloading with a custom URL:

```
$ pip install https://github.com/explosion/spacy-models/
releases/download/en_core_web_lg-2.0.0/en_core_web_
lg-2.0.0.tar.gz
```

You can install a local file as follows:

```
$ pip install /Users/yourself/en_core_web_lg-2.0.0.tar.gz
```

This installs the model into your `site-packages` directory. Then we run `spacy.load()` to load the model via its package name, create a shortcut link to give it a custom name (usually a shorter name), or import it as a module.

Importing the language model as a module is also possible:

```
import en_core_web_md
nlp = en_core_web_md.load()
doc = nlp('I have a ginger cat.')
```

> **Pro tip**
> In professional software development, we usually download models as part of an automated pipeline. In this case, it's not feasible to use spaCy's `download` command; rather, we use `pip` with the model URL. You can add the model into your `requirements.txt` file as a package as well.

How you like to load your models is your choice and also depends on the project requirements you're working on.

At this point, we're ready to explore the spaCy world. Let's now learn about spaCy's powerful visualization tool, **displaCy**.

Visualization with displaCy

Visualization is an important tool that should be in every data scientist's toolbox. Visualization is the easiest way to explain some concepts to your colleagues, your boss, and any technical or non-technical audience. Visualization of language data is specifically useful and allows you to identify patterns in your data at a glance.

There are many Python libraries and plugins such as *Matplotlib*, *seaborn*, *TensorBoard*, and so on. Being an industrial library, spaCy comes with its own visualizer – **displaCy**. In this subsection, you'll learn how to spin up a displaCy server on your machine, in a Jupyter notebook, and in a web application. You'll also learn how to export the graphics you created as an image file, customize your visualizations, and make manual annotations without creating a `Doc` object. We'll start by exploring the easiest way – using displaCy's interactive demo.

Getting started with displaCy

Go ahead and navigate to `https://explosion.ai/demos/displacy` to use the interactive demo. Enter your text in the **Text to parse** box and then click the search icon on the right to generate the visualization. The result might look like the following:

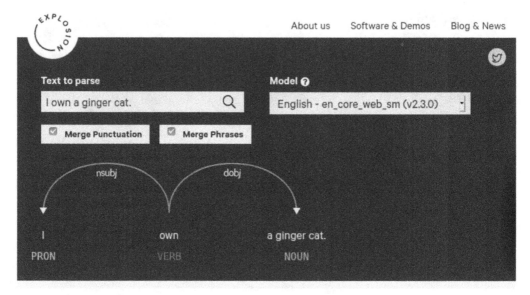

Figure 1.11 – displaCy's online demo

The visualizer performs two syntactic parses, POS tagging, and a **dependency parse**, on the submitted text to visualize the sentence's syntactic structure. Don't worry about how POS tagging and dependency parsing work, as we'll explore them in the upcoming chapters. For now, just think of the result as a sentence structure.

You'll notice two tick boxes, **Merge Punctuation** and **Merge Phrases**. Merging punctuation merges the punctuation tokens into the previous token and serves a more compact visualization (it works like a charm on long documents).

The second option, **Merge Phrases**, again gives more compact dependency trees. This option merges adjectives and nouns into one phrase; if you don't merge, then adjectives and nouns will be displayed individually. This feature is useful for visualizing long sentences with many noun phrases. Let's see the difference with an example sentence: `They were beautiful and healthy kids with strong appetites.` It contains two noun phrases, `beautiful and healthy kids` and `strong appetite`. If we merge them, the result is as follows:

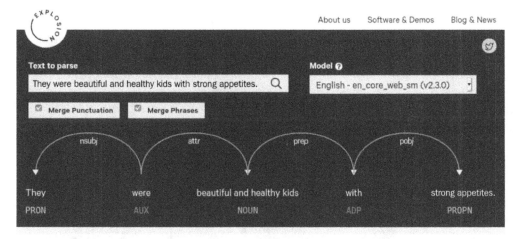

Figure 1.12 – An example parse with noun phrases merged

Without merging, every adjective and noun are shown individually:

Figure 1.13 – A parse of the same sentence, unmerged

The second parse is a bit too cumbersome and difficult to read. If you work on a text with long sentences such as law articles or Wikipedia entries, we definitely recommend merging.

You can choose a statistical model from the **Model** box on the right for the currently supported languages. This option allows you to play around with the language models without having to download and install them on your local machine.

Entity visualizer

displaCy's entity visualizer highlights the named entities in your text. The online demo lives at `https://explosion.ai/demos/displacy-ent/`. We didn't go through named entities yet, but you can think of them as proper nouns for important entities such as people's names, company names, dates, city and country names, and so on. Extracting entities will be covered in *Chapter 3*, *Linguistic Features*, and *Chapter 4*, *Rule-Based Matching*, in detail.

The online demo works similar to the syntactic parser demo. Enter your text into the textbox and hit the search button. Here is an example:

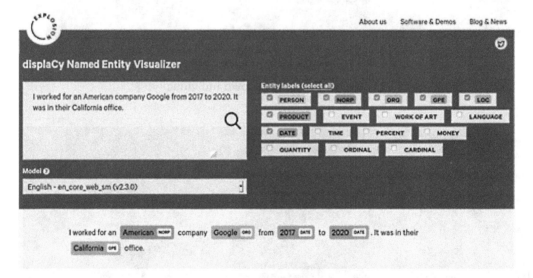

Figure 1.14 – An example entity visualization

The right side contains tick boxes for entity types. You can tick the boxes that match your text type such as, for instance, **MONEY** and **QUANTITY** for a financial text. Again, just like in the syntactic parser demo, you can choose from the available models.

Visualizing within Python

With the introduction of the latest version of spaCy, the displaCy visualizers are integrated into the core library. This means that you can start using displaCy immediately after installing spaCy on your machine! Let's go through some examples.

The following code segment is the easiest way to spin up displaCy on your local machine:

```
import spacy
from spacy import displacy
nlp = spacy.load('en_core_web_md')
doc= nlp('I own a ginger cat.')
displacy.serve(doc, style='dep')
```

As you can see from the preceding snippet, the following is what we did:

1. We import spaCy.

2. Following that, we import displaCy from the core library.

3. We load the English model that we downloaded in the *Installing spaCy's statistical models* section.

4. Once it is loaded, we create a Doc object to pass to displaCy.

5. We then started the displaCy web server via calling serve().

6. We also passed dep to the style parameter to see the dependency parsing result.

After firing up this code, you should see a response from displaCy as follows:

```
Using the 'dep' visualizer
Serving on http://0.0.0.0:5000 ...

127.0.0.1 - - [08/Oct/2020 15:25:15] "GET / HTTP/1.1" 200 4221
```

Figure 1.15 – Firing up displaCy locally

The response is added along with a link, `http://0.0.0.0:5000`, this is the local address where displaCy renders your graphics. Please click the link and navigate to the web page. You should see the following:

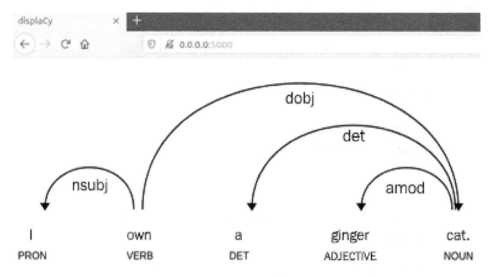

Figure 1.16 – View the result visualization in your browser

This means that displaCy generated a dependency parse result visualization and rendered it on your localhost. After you're finished with displaying the visual and you want to shut down the server, you can press *Ctrl +C* to shut down the displaCy server and go back to the Python shell:

```
Using the 'dep' visualizer
Serving on http://0.0.0.0:5000 ...

27.0.0.1 - - [08/Oct/2020 15:25:15] "GET / HTTP/1.1" 200 4221
CShutting down server on port 5000.
>>
```

Figure 1.17 – Shutting down the displaCy server

After shutting down, you won't be able to visualize more examples, but you'll continue seeing the results you already generated.

If you wish to use another port or you get an error because the port 5000 is already in use, you can use the `port` parameter of displaCy with another port number. Replacing the last line of the preceding code block with the following line will suffice:

```
display.serve(doc, style='dep', port= '5001')
```

Here, we provide the port number 5001 explicitly. In this case, displaCy will render the graphics on http://0.0.0.0:5001.

Generating an entity recognizer is done similarly. We pass ent to the style parameter instead of dep:

```
import spacy
from spacy import displacy
nlp = spacy.load('en_core_web_md')
doc= nlp('Bill Gates is the CEO of Microsoft.')
displacy.serve(doc, style='ent')
```

The result should look like the following:

Figure 1.18 – The entity visualization is displayed on your browser

Let's move on to other platforms we can use for displaying the results.

Using displaCy in Jupyter notebooks

Jupyter notebook is an important part of daily data science work. Fortunately, displaCy can spot whether you're currently coding in a Jupyter notebook environment and returns markup that can be directly displayed in a cell.

If you don't have Jupyter notebook installed on your system but wish to use it, you can follow the instructions at https://test-jupyter.readthedocs.io/en/latest/install.html.

This time we'll call `render()` instead of `serve()`. The rest of the code is the same. You can type/paste the following code into your Jupyter notebook:

```
import spacy
from spacy import displacy
nlp = spacy.load('en_core_web_md')
doc= nlp('Bill Gates is the CEO of Microsoft.')
displacy.render(doc, style='dep')
```

The result should look like the following:

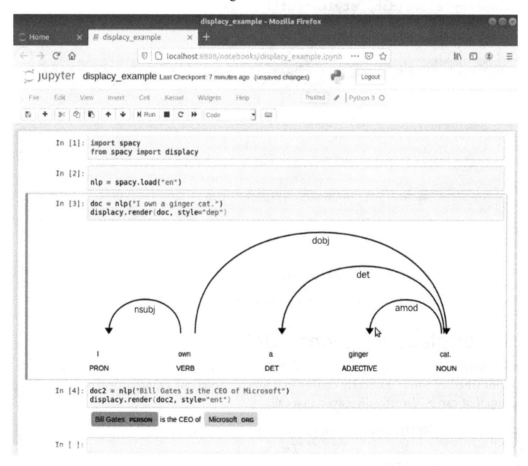

Figure 1.19 – displaCy rendering results in a Jupyter notebook

Exporting displaCy graphics as an image file

Often, we need to export the graphics that we generated with displaCy as image files to place them into presentations, articles, or papers. We can call displaCy in this case as well:

```
import spacy
from spacy import displacy
from pathlib import Path
nlp = spacy.load('en_core_web_md')
doc = nlp('I'm a butterfly.')
svg = display.render(doc, style='dep', jupyter=False)
filename = 'butterfly.svg'
  output_path = Path ('/images/' + file_name)
output_path.open('w', encoding='utf-8').write(svg)
```

We import spaCy and displaCy. We load the English language model, then create a Doc object as usual. Then we call display.render() and capture the output to the svg variable. The rest is writing the svg variable to a file called butterfly.svg.

We have reached the end of the visualization chapter here. We created good-looking visuals and learned the details of creating visuals with displaCy. If you wish to find out how to use different background images, background colors, and fonts, you can visit the displaCy documentation at http://spacy.io/usage/visualizers.

Often, we need to create visuals with different colors and styling, and the displaCy documentation contains detailed information about styling. The documentation also includes how to embed displaCy into your web applications. spaCy is well documented as a project and the documentation contains everything we need!

Summary

This chapter gave you an introduction to NLP with Python and spaCy. You now have a brief idea about why to use Python for language processing and the reasons to prefer spaCy for creating your NLP applications. We also got started on our spaCy journey by installing spaCy and downloading language models. This chapter also introduced us to the visualization tool – displaCy.

In the next chapter, we will continue our exciting spaCy journey with spaCy core operations such as tokenization and lemmatization. It'll be our first encounter with spaCy features in detail. Let's go ahead and explore more together!

2
Core Operations with spaCy

In this chapter, you will learn the core operations with spaCy, such as creating a language pipeline, tokenizing the text, and breaking the text into its sentences.

First, you'll learn what a language processing pipeline is and the pipeline components. We'll continue with general spaCy conventions – important classes and class organization – to help you to better understand spaCy library organization and develop a solid understanding of the library itself.

You will then learn about the first pipeline component – **Tokenizer**. You'll also learn about an important linguistic concept – **lemmatization** – along with its applications in **natural language understanding** (NLU). Following that, we will cover **container classes** and **spaCy data structures** in detail. We will finish the chapter with useful spaCy features that you'll use in everyday NLP development.

We're going to cover the following main topics in this chapter:

- Overview of spaCy conventions
- Introducing tokenization
- Understanding lemmatization
- spaCy container objects
- More spaCy features

Technical requirements

The chapter code can be found at the book's GitHub repository: `https://github.com/PacktPublishing/Mastering-spaCy/tree/main/Chapter02`

Overview of spaCy conventions

Every NLP application consists of several steps of processing the text. As you can see in the first chapter, we have always created instances called `nlp` and `doc`. But what did we do exactly?

When we call `nlp` on our text, spaCy applies some processing steps. The first step is tokenization to produce a `Doc` object. The `Doc` object is then processed further with a **tagger**, a **parser**, and an **entity recognizer**. This way of processing the text is called a **language processing pipeline**. Each pipeline component returns the processed `Doc` and then passes it to the next component:

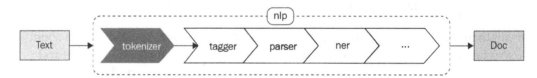

Figure 2.1 – A high-level view of the processing pipeline

A spaCy pipeline object is created when we load a language model. We load an English model and initialize a pipeline in the following code segment:

```
import spacy
nlp = spacy.load("en_core_web_md")
doc = nlp("I went there")
```

What happened exactly in the preceding code is as follows:

1. We started by importing `spaCy`.

2. In the second line, `spacy.load()` returned a `Language` class instance, `nlp`. The `Language` class is *the text processing pipeline*.

3. After that, we applied `nlp` on the sample sentence `I went there` and got a `Doc` class instance, `doc`.

The `Language` class applies all of the preceding pipeline steps to your input sentence behind the scenes. After applying `nlp` to the sentence, the `Doc` object contains tokens that are tagged, lemmatized, and marked as entities if the token is an entity (we will go into detail about what are those and how it's done later). Each pipeline component has a well-defined task:

NAME	COMPONENT	CREATES	DESCRIPTION
tokenizer	Tokenizer	Doc	Segment text into tokens.
tagger	Tagger	Doc[i].tag	Assign part-of-speech tags.
parser	DependencyParser	Doc[i].head, Doc[i].dep, Doc.sents, Doc.noun_chunks	Assign dependency labels.
ner	EntityRecognizer	Doc.ents, Doc[i].ent_iob, Doc[i].ent_type	Detect and label named entities.

Figure 2.2 – Pipeline components and tasks

The spaCy language processing pipeline always *depends on the statistical model* and its capabilities. This is why we always load a language model with `spacy.load()` as the first step in our code.

Each component corresponds to a `spaCy` class. `spaCy` classes have self-explanatory names such as **Language**, **Doc**, and **Vocab**. We already used `Language` and `Doc` classes – let's see all of the processing pipeline classes and their duties:

Processing pipeline

TYPE	DESCRIPTION
Language	A text-processing pipeline. Usually you'll load this once per process as nlp and pass the instance around your application.
Tokenizer	Segment text, and create Doc objects with the discovered segment boundaries.
Lemmatizer	Determine the base forms of words.
Morphology	Assign linguistic features like lemmas, noun case, verb tense etc. based on the word and its part-of-speech tag.
Tagger	Annotate part-of-speech tags on Doc objects.
DependencyParser	Annotate syntactic dependencies on Doc objects.
EntityRecognizer	Annotate named entities, e.g. persons or products, on Doc objects.
TextCategorizer	Assign categories or labels to Doc objects.
Matcher	Match sequences of tokens, based on pattern rules, similar to regular expressions.
PhraseMatcher	Match sequences of tokens, based on phrases.
EntityRuler	Add entity spans to the Doc using token-based rules or exact phrase matches.
Sentencizer	Implement custom sentence boundary detection logic that doesn't require the dependency parse.

Figure 2.3 – spaCy processing pipeline classes

Don't be intimated by the number of classes; each class has unique features to help you process your text better.

There are more data structures to represent text data and language data. Container classes such as Doc hold information about sentences, words, and the text. There are also container classes other than Doc:

Container objects

NAME	DESCRIPTION
Doc	A container for accessing linguistic annotations.
Span	A slice from a Doc object.
Token	An individual token - i.e. a word, puntuation symbol, whitespace, etc.
Lexeme	An entry in the vocabulary. It's a word type with no context, as opposed to a word token. It therefore has no part-of-speech tag, dependency parse etc.

Figure 2.4 – spaCy container classes

Finally, spaCy provides helper classes for vectors, language vocabulary, and annotations. We'll see the Vocab class often in this book. Vocab represents a language's vocabulary. Vocab contains all the words of the language model we loaded:

Other classes

NAME	DESCRIPTION
Vocab	A lookup table for the vocabulary that allows you to access Lexeme objects.
StringStore	Map strings to and from hash values.
Vectors	Container class for vector data keyed by string.
GoldParse	Collection for training annotations.
GoldCorpus	An annotated corpus, using the JSON file format. Manages annotations for tagging, dependency parsing and NER.

Figure 2.5 – spaCy helper classes

The spaCy library's backbone data structures are Doc and Vocab. The Doc object abstracts the text by owning the sequence of tokens and all their properties. The Vocab object provides a centralized set of strings and lexical attributes to all the other classes. This way spaCy avoids storing multiple copies of linguistic data:

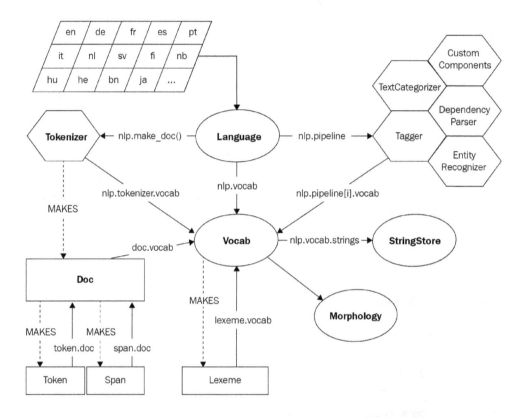

Figure 2.6 – spaCy architecture

You can divide the objects composing the preceding spaCy architecture into two: **containers** and **processing pipeline components**. In this chapter, we'll first learn about two basic components, **Tokenizer** and **Lemmatizer**, then we'll explore **Container** objects further.

spaCy does all these operations for us behind the scenes, allowing us to concentrate on our own application's development. With this level of abstraction, using spaCy for NLP application development is no coincidence. Let's start with the Tokenizer class and see what it offers for us; then we will explore all the container classes one by one throughout the chapter.

Introducing tokenization

We saw in *Figure 2.1* that the first step in a text processing pipeline is tokenization. Tokenization is always the first operation because all the other operations require the tokens.

Tokenization simply means splitting the sentence into its tokens. A **token** is a unit of semantics. You can think of a token as the smallest meaningful part of a piece of text. Tokens can be words, numbers, punctuation, currency symbols, and any other meaningful symbols that are the building blocks of a sentence. The following are examples of tokens:

- USA
- N.Y.
- city
- 33
- 3rd
- !
- ...
- ?
- 's

Input to the spaCy tokenizer is a Unicode text and the result is a Doc object. The following code shows the tokenization process:

```
import spacy
nlp = spacy.load("en_core_web_md")
doc = nlp("I own a ginger cat.")
print ([token.text for token in doc])
['I', 'own', 'a', 'ginger', 'cat', '.']
```

The following is what we just did:

1. We start by importing spaCy.
2. Then we loaded the English language model via the en shortcut to create an instance of the nlp Language class.

3. Next, we apply the `nlp` object to the input sentence to create a `Doc` object, `doc`. A `Doc` object is a container for a sequence of `Token` objects. spaCy generates the `Token` objects implicitly when we created the `Doc` object.

4. Finally, we print a list of the preceding sentence's tokens.

That's it, we made the tokenization with just three lines of code. You can visualize the tokenization with indexing as follows:

0	1	2	3	4	5
I	own	a	ginger	cat	.

Figure 2.7 – Tokenization of "I own a ginger cat."

As the examples suggest, tokenization can indeed be tricky. There are many aspects we should pay attention to: punctuation, whitespaces, numbers, and so on. Splitting from the whitespaces with `text.split(" ")` might be tempting and looks like it is working for the example sentence *I own a ginger cat.*

How about the sentence `"It's been a crazy week!!!"`? If we make a `split(" ")` the resulting tokens would be `It's`, `been`, `a`, `crazy`, `week!!!`, which is not what you want. First of all, `It's` is not one token, it's two tokens: `it` and `'s`. `week!!!` is not a valid token as the punctuation is not split correctly. Moreover, `!!!` should be tokenized per symbol and should generate three *!*'s. (This may not look like an important detail, but trust me, it is important for *sentiment analysis*. We'll cover sentiment analysis in *Chapter 8, Text Classification with spaCy*.) Let's see what spaCy tokenizer has generated:

```
import spacy
nlp = spacy.load("en_core_web_md")
doc = nlp("It's been a crazy week!!!")
print ([token.text for token in doc])
['It', "'s", 'been', 'a', 'crazy', 'week', '!', '!', '!']
```

This time the sentence is split as follows:

0	1	2	3	4	5	6	7	8
It	's	been	a	crazy	week	!	!	!

Figure 2.8 – Tokenization of apostrophe and punctuations marks

How does spaCy know where to split the sentence? Unlike other parts of the pipeline, the tokenizer doesn't need a statistical model. Tokenization is based on language-specific rules. You can see examples the language specified data here: `https://github.com/explosion/spaCy/tree/master/spacy/lang`.

Tokenizer exceptions define rules for exceptions, such as `it's`, `don't`, `won't`, abbreviations, and so on. If you look at the rules for English: `https://github.com/explosion/spaCy/blob/master/spacy/lang/en/tokenizer_exceptions.py`, you will see that rules look like `{ORTH: "n't", LEMMA: "not"}`, which describes the splitting rule for `n't` to the tokenizer.

The prefixes, suffixes, and infixes mostly describe how to deal with punctuation – for example, we split at a period if it is at the end of the sentence, otherwise, most probably it's part of an abbreviation such as N.Y. and we shouldn't touch it. Here, `ORTH` means the text and `LEMMA` means the base word form without any inflections. The following example shows you the execution of the spaCy tokenization algorithm:

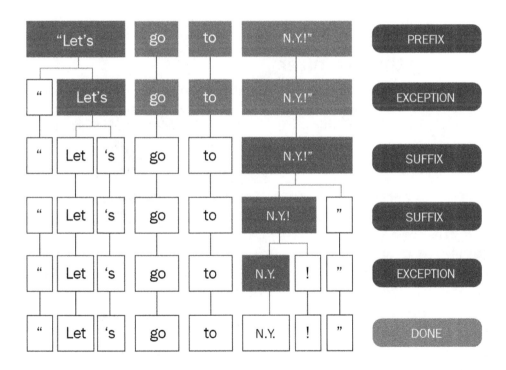

Figure 2.9 – spaCy performing tokenization with exception rules (image taken from spaCy tokenization guidelines (`https://spacy.io/usage/linguistic-features#tokenization`))

Tokenization rules depend on the grammatical rules of the individual language. Punctuation rules such as splitting periods, commas, or exclamation marks are more or less similar for many languages; however, some rules are specific to the individual language, such as abbreviation words and apostrophe usage. spaCy supports each language having its own specific rules by allowing hand-coded data and rules, as each language has its own subclass.

> **Tip**
>
> spaCy provides non-destructive tokenization, which means that we always will be able to recover the original text from the tokens. Whitespace and punctuation information is preserved during tokenization, so the input text is preserved as it is.

Every Language object contains a Tokenizer object. The Tokenizer class is the class that performs the tokenization. You don't often call this class directly when you create a Doc class instance, while Tokenizer class acts behind the scenes. When we want to customize the tokenization, we need to interact with this class. Let's see how it is done.

Customizing the tokenizer

When we work with a specific domain such as medicine, insurance, or finance, we often come across words, abbreviations, and entities that needs special attention. Most domains that you'll process have characteristic words and phrases that need custom tokenization rules. Here's how to add a special case rule to an existing Tokenizer class instance:

```
import spacy
from spacy.symbols import ORTH
nlp = spacy.load("en_core_web_md")
doc = nlp("lemme that")
print([w.text for w in doc])
['lemme', 'that']
special_case = [{ORTH: "lem"}, {ORTH: "me"}]
nlp.tokenizer.add_special_case("lemme", special_case)
print([w.text for w in nlp("lemme that")])
['lem', 'me', 'that']
```

Here is what we did:

1. We again started by importing spacy.

2. Then, we imported the ORTH symbol, which means orthography; that is, text.

3. We continued with creating a `Language` class object, `nlp`, and created a Doc object, `doc`.

4. We defined a special case, where the word `lemme` should tokenize as two tokens, `lem` and `me`.

5. We added the rule to the `nlp` object's tokenizer.

6. The last line exhibits how the fresh rule works.

When we define custom rules, punctuation splitting rules will still apply. Our special case will be recognized as a result, even if it's surrounded by punctuation. The tokenizer will divide punctuation step by step, and apply the same process to the remaining substring:

```
print([w.text for w in nlp("lemme!")])
['lem', 'me', '!']
```

If you define a special case rule with punctuation, the special case rule will take precedence over the punctuation splitting:

```
nlp.tokenizer.add_special_case("...lemme...?", [{"ORTH": "...
lemme...?"}])
print([w.text for w in nlp("...lemme...?")])
'...lemme...?'
```

> **Pro tip**
> Modify the tokenizer by adding new rules only if you really need to. Trust me, you can get quite unexpected results with custom rules. One of the cases where you really need it is when working with Twitter text, which is usually full of hashtags and special symbols. If you have social media text, first feed some sentences into the spaCy NLP pipeline and see how the tokenization works out.

Debugging the tokenizer

The spaCy library has a tool for debugging: `nlp.tokenizer.explain(sentence)`. It returns (`tokenizer rule/pattern, token`) **tuples** to help us understand what happened exactly during the tokenization. Let's see an example:

```
import spacy
nlp = spacy.load("en_core_web_md")
text = "Let's go!"
doc = nlp(text)
```

```
tok_exp = nlp.tokenizer.explain(text)
for t in tok_exp:
    print(t[1], "\t", t[0])
Let     SPECIAL-1
's      SPECIAL-2
go      TOKEN
!       SUFFIX
```

In the preceding code, we imported spacy and created a Language class instance, nlp, as usual. Then we created a Doc class instance with the sentence Let's go!. After that, we asked the Tokenizer class instance, tokenizer, of nlp for an explanation of the tokenization of this sentence. nlp.tokenizer.explain() explained the rules that the tokenizer used one by one.

After splitting a sentence into its tokens, it's time to split a text into its sentences.

Sentence segmentation

We saw that breaking a sentence into its tokens is not a straightforward task at all. How about breaking a text into sentences? It's indeed a bit more complicated to mark where a sentence starts and ends due to the same reasons of punctuation, abbreviations, and so on.

A Doc object's sentences are available via the doc.sents property:

```
import spacy
nlp = spacy.load("en_core_web_md")
text = "I flied to N.Y yesterday. It was around 5 pm."
doc = nlp(text)
for sent in doc.sents:
    print(sent.text)
I flied to N.Y yesterday.
It was around 5 pm.
```

Determining sentence boundaries is a more complicated task than tokenization. As a result, spaCy uses the dependency parser to perform sentence segmentation. This is a unique feature of spaCy – no other library puts such a sophisticated idea into practice. The results are very accurate in general, unless you process text of a very specific genre, such as from the conversation domain, or social media text.

Now we know how to segment a text into sentences and tokenize the sentences. We're ready to process the tokens one by one. Let's start with lemmatization, a commonly used operation in semantics including sentiment analysis.

Understanding lemmatization

A **lemma** is the base form of a token. You can think of a lemma as the form in which the token appears in a dictionary. For instance, the lemma of *eating* is *eat*; the lemma of *eats* is *eat*; *ate* similarly maps to *eat*. Lemmatization is the process of reducing the word forms to their lemmas. The following code is a quick example of how to do lemmatization with spaCy:

```
import spacy
nlp = spacy.load("en_core_web_md")
doc = nlp("I went there for working and worked for 3 years.")
for token in doc:
    print(token.text, token.lemma_)
I -PRON-
went go
there there
for for
working work
and and
worked work
for for
3 3
years year
. .
```

By now, you should be familiar with what the first three lines of the code do. Recall that we import the `spacy` library, load an English model using `spacy.load`, create a pipeline, and apply the pipeline to the preceding sentence to get a Doc object. Here we iterated over tokens to get their text and lemmas.

In the first line you see `-PRON-`, which doesn't look like a real token. This is a **pronoun lemma**, a special token for lemmas of personal pronouns. This is an exception for semantic purposes: the personal pronouns *you, I, me, him, his,* and so on look different, but in terms of meaning, they're in the same group. spaCy offers this trick for the pronoun lemmas.

No worries if all of this sounds too abstract – let's see lemmatization in action with a real-world example.

Lemmatization in NLU

Lemmatization is an important step in NLU. We'll go over an example in this subsection. Suppose that you design an NLP pipeline for a ticket booking system. Your application processes a customer's sentence, extracts necessary information from it, and then passes it to the booking API.

The NLP pipeline wants to extract the form of the travel (a flight, bus, or train), the destination city, and the date. The first thing the application needs to verify is the means of travel:

```
fly - flight - airway - airplane - plane
bus
railway - train
```

We have this list of keywords and we want to recognize the means of travel by searching the tokens in the keywords list. The most compact way of doing this search is by looking up the token's lemma. Consider the following customer sentences:

```
List me all flights to Atlanta.
I need a flight to NY.
I flew to Atlanta yesterday evening and forgot my baggage.
```

Here, we don't need to include all word forms of the verb *fly* (*fly, flying, flies, flew,* and *flown*) in the keywords list and similar for the word `flight`; we reduced all possible variants to the base forms – *fly* and *flight*. Don't think of English only; languages such as Spanish, German, and Finnish have many word forms from a single lemma as well.

Lemmatization also comes in handy when we want to recognize the destination city. There are many nicknames available for global cities and the booking API can process only the official names. The default tokenizer and lemmatizer won't know the difference between the official name and the nickname. In this case, you can add special rules, as we saw in the *Introducing tokenization* section. The following code plays a small trick:

```
import spacy
nlp  = spacy.load("en_core_web_md")
nlp.get_pipe("attribute_ruler").add([[{"TEXT": "Angeltown"}]],
{"LEMMA": "Los Angeles"})
doc = nlp("I am flying to Angeltown")
```

```
for token in doc:
    print(token.text, token.lemma_)
I -PRON-
am be
flying fly
to to
Angeltown Los Angeles
```

We defined a special case for the word `Angeltown` by replacing its lemma with the official name `Los Angeles`. Then we added this special case to the `AttributeRuler` component. When we print the token lemmas, we see that `Angeltown` maps to `Los Angeles` as we wished.

Understanding the difference between lemmatization and stemming

A lemma is the base form of a word and is always a member of the language's vocabulary. The stem does not have to be a valid word at all. For instance, the lemma of *improvement* is *improvement*, but the stem is *improv*. You can think of the stem as the smallest part of the word that carries the meaning. Compare the following examples:

Word	Lemma
university	university
universe	universe
universal	universal
universities	university
universes	universe
improvement	improvement
improvements	improvements
improves	improve

The preceding word-lemma examples show how lemma is calculated by following the grammatical rules of the language. Here, the lemma of a plural form is the singular form, and the lemma of a third-person verb is the base form of the verb. Let's compare them to the following examples of word-stem pairs:

Word	Stem
university	univers
universe	univer

universal	univers
universities	universi
universes	univers
improvement	improv
improvements	improv
improves	improv

The first and the most important point to notice in the preceding examples is that the stem does not have to be a valid word in the language. The second point is that many words can map to the same stem. Also, words from different grammatical categories can map to the same stem; here for instance, the noun *improvement* and the verb *improves* both map to *improv*.

Though stems are not valid words, they still carry meaning. That's why stemming is commonly used in NLU applications.

Stemming algorithms don't know anything about the grammar of the language. This class of algorithms works rather by trimming some common suffixes and prefixes from the beginning or end of the word.

Stemming algorithms are rough, they cut the word from head and tail. There are several stemming algorithms available for English, including Porter and Lancaster. You can play with different stemming algorithms on NLTK's demo page at `http://text-processing.com/demo/stem/`.

Lemmatization, on the other hand, takes the morphological analysis of the words into consideration. To do so, it is important to obtain the dictionaries for the algorithm to look through in order to link the form back to its lemma.

spaCy provides lemmatization via dictionary lookup and each language has its own dictionary.

> **Tip**
> Both stemming and lemmatization have their own advantages. Stemming gives very good results if you apply only statistical algorithms to the text, without further semantic processing such as pattern lookup, entity extraction, coreference resolution, and so on. Also stemming can trim a big corpus to a more moderate size and give you a compact representation. If you also use linguistic features in your pipeline or make a keyword search, include lemmatization. Lemmatization algorithms are accurate but come with a cost in terms of computation.

spaCy container objects

At the beginning of this chapter, we saw a list of container objects including **Doc**, **Token**, **Span**, and **Lexeme**. We already used Token and Doc in our code. In this subsection, we'll see the properties of the container objects in detail.

Using container objects, we can access the linguistic properties that spaCy assigns to the text. A container object is a logical representation of the text units such as a document, a token, or a slice of the document.

Container objects in spaCy follow the natural structure of the text: a document is composed of sentences and sentences are composed of tokens.

We most widely use Doc, Token, and Span objects in development, which represent a document, a single token, and a phrase, respectively. A container can contain other containers, for instance a document contains tokens and spans.

Let's explore each class and its useful properties one by one.

Doc

We created Doc objects in our code to represent the text, so you might have already figured out that Doc represents a text.

We already know how to create a Doc object:

```
doc = nlp("I like cats.")
```

doc.text returns a Unicode representation of the document text:

```
doc.text
I like cats.
```

The building block of a Doc object is Token, hence when you iterate a Doc you get Token objects as items:

```
for token in doc:
    print(token.text)
I
like
cats
.
```

The same logic applies to indexing:

```
doc[1]
like
```

The length of a Doc is the number of tokens it includes:

```
len(doc)
4
```

We already saw how to get the text's sentences. doc.sents returns an iterator to the list of sentences. Each sentence is a Span object:

```
doc = nlp("This is a sentence. This is the second sentence")
doc.sents
<generator object at 0x7f21dc565948>
sentences = list(doc.sents)
sentences
["This is a sentence.", "This is the second sentence."]
```

doc.ents gives named entities of the text. The result is a list of Span objects. We'll see named entities in detail later – for now, think of them as proper nouns:

```
doc = nlp("I flied to New York with Ashley.")
doc.ents
(New York, Ashley)
```

Another syntactic property is doc.noun_chunks. It yields the noun phrases found in the text:

```
doc = nlp("Sweet brown fox jumped over the fence.")
list(doc.noun_chunks)
[Sweet brown fox, the fence]
```

doc.lang_ returns the language that doc created:

```
doc.lang_
'en'
```

A useful method for serialization is `doc.to_json`. This is how to convert a Doc object to JSON:

```
doc = nlp("Hi")
json_doc = doc.to_json()
{
  "text": "Hi",
  "ents": [],
  "sents": [{"start": 0, "end": 3}],
  "tokens": [{"id": 0, "start": 0, "end": 3, "pos": "INTJ",
"tag": "UH", "dep": "ROOT", "head": 0}]
}
```

> **Pro tip**
> You might have noticed that we call `doc.lang_`, not `doc.lang`. `doc.lang` returns the language ID, whereas `doc.lang_` returns the Unicode string of the language, that is, the name of the language. You can see the same convention with Token features in the following, for instance, `token.lemma_`, `token.tag_`, and `token.pos_`.

The Doc object has very useful properties with which you can understand a sentence's syntactic properties and use them in your own applications. Let's move on to the Token object and see what it offers.

Token

A Token object represents a word. Token objects are the building blocks of Doc and Span objects. In this section, we will cover the following properties of the `Token` class:

- `token.text`
- `token.text_with_ws`
- `token.i`
- `token.idx`
- `token.doc`
- `token.sent`
- `token.is_sent_start`
- `token.ent_type`

We usually don't construct a Token object directly, rather we construct a Doc object then access its tokens:

```
doc = nlp("Hello Madam!")
doc[0]
Hello
```

`token.text` is similar to `doc.text` and provides the underlying Unicode string:

```
doc[0].text
Hello
```

`token.text_with_ws` is a similar property. It provides the text with a trailing whitespace if present in the `doc`:

```
doc[0].text_with_ws
'Hello '
doc[2].text_with_ws
'!"
```

Finding the length of a token is similar to finding the length of a Python string:

```
len(doc[0])
5
```

`token.i` gives the index of the token in `doc`:

```
token = doc[2]
token.i
2
```

`token.idx` provides the token's character offset (the character position) in `doc`:

```
doc[0].idx
0
doc[1].idx
6
```

We can also access the doc that created the token as follows:

```
token = doc[0]
token.doc
Hello Madam!
```

Getting the sentence that the token belongs to is done in a similar way to accessing the doc that created the token:

```
token = doc[1]
token.sent
Hello Madam!
```

token.is_sent_start is another useful property; it returns a Boolean indicating whether the token starts a sentence:

```
doc = nlp("He entered the room. Then he nodded.")
doc[0].is_sent_start
True
doc[5].is_sent_start
True
doc[6].is_sent_start
False
```

These are the basic properties of the Token object that you'll use every day. There is another set of properties that are more related to syntax and semantics. We already saw how to calculate the token lemma in the previous section:

```
doc = nlp("I went there.")
doc[1].lemma_
'go'
```

You already learned that doc.ents gives the named entities of the document. If you want to learn what sort of entity the token is, use token.ent_type_:

```
doc = nlp("President Trump visited Mexico City.")
doc.ents
(Trump, Mexico City)
doc[1].ent_type_
'PERSON'
doc[3].ent_type_
```

```
'GPE'  # country, city, state
 doc[4].ent_type_
'GPE'  # country, city, state
 doc[0].ent_type_
 ''  # not an entity
```

Two syntactic features related to POS tagging are `token.pos_` and `token.tag`. We'll learn what they are and how to use them in the next chapter.

Another set of syntactic features comes from the dependency parser. These features are `dep_`, `head_`, `conj_`, `lefts_`, `rights_`, `left_edge_`, and `right_edge_`. We'll cover them in the next chapter as well.

> **Tip**
>
> It is totally normal if you don't remember all the features afterward. If you don't remember the name of a feature, you can always do `dir(token)` or `dir(doc)`. Calling `dir()` will print all the features and methods available on the object.

The Token object has a rich set of features, enabling us to process the text from head to toe. Let's move on to the Span object and see what it offers for us.

Span

Span objects represent phrases or segments of the text. Technically, a Span has to be a contiguous sequence of tokens. We usually don't initialize Span objects, rather we slice a Doc object:

```
doc = nlp("I know that you have been to USA.")
 doc[2:4]
"that you"
```

Trying to slice an invalid index will raise an `IndexError`. Most indexing and slicing rules of Python strings are applicable to Doc slicing as well:

```
doc = nlp("President Trump visited Mexico City.")
 doc[4:]  # end index empty means rest of the string
City.
 doc[3:-1]  # minus indexes are supported
 doc[6:]
```

```
Traceback (most recent call last):
  File "<stdin>", line 1, in <module>
  File "span.pyx", line 166, in spacy.tokens.span.Span.__repr__
  File "span.pyx", line 503, in spacy.tokens.span.Span.text.__
get__
  File "span.pyx", line 190, in spacy.tokens.span.Span.__
getitem__
IndexError: [E201] Span index out of range.
  doc[1:1]  # empty spans are not allowed
Traceback (most recent call last):
  File "<stdin>", line 1, in <module>
  File "span.pyx", line 166, in spacy.tokens.span.Span.__repr__
  File "span.pyx", line 503, in spacy.tokens.span.Span.text.__
get__
  File "span.pyx", line 190, in spacy.tokens.span.Span.__
getitem__
IndexError: [E201] Span index out of range.
```

There is one more way to create a Span – we can make a character-level slice of a Doc object with char_span :

```
doc = nlp("You love Atlanta since you're 20.")
doc.char_span(4, 16)
love Atlanta
```

The building blocks of a Span object are Token objects. If you iterate over a Span object you get Token objects:

```
doc = nlp("You went there after you saw me")
span = doc[2:4]
for token in span:
    print(token)
there
after
```

You can think of the Span object as a *junior* Doc object, indeed it's a view of the Doc object it's created from. Hence most of the features of Doc are applicable to Span as well. For instance, `len` is identical:

```
doc = nlp("Hello Madam!")
span = doc[1:2]
len(span)
1
```

Span object also supports indexing. The result of slicing a Span object is another Span object:

```
doc = nlp("You went there after you saw me")
span = doc[2:6]
span
there after you saw
subspan = span[1:3]
after you
```

`char_spans` also works on Span objects. Remember the Span class is a junior Doc class, so we can create character-indexed spans on Span objects as well:

```
doc = nlp("You went there after you saw me")
span = doc[2:6]
span.char_span(15,24)
after you
```

Just like a Token knows the Doc object it's created from; Span also knows the Doc object it's created from:

```
doc = nlp("You went there after you saw me")
span = doc[2:6]
span.doc
You went there after you saw me
span.sent
You went there after you saw me
```

We can also locate the Span in the original Doc:

```
doc = nlp("You went there after you saw me")
span = doc[2:6]
span.start
2
span.end
6
span.start_char
9
span.end_char
28
```

span.start is the index of the first token of the Span and span.start_char is the start offset of the Span at character level.

If you want a brand-new Doc object, you can call span.as_doc(). It copies the data into a new Doc object:

```
doc = nlp("You went there after you saw me")
span = doc[2:6]
type(span)
<class 'spacy.tokens.span.Span'>
small_doc = span.as_doc()
type(small_doc)
<class 'spacy.tokens.doc.Doc'>
```

span.ents, span.sent, span.text, and span.text_wth_ws are similar to their corresponding Doc and Token methods.

Dear readers, we have reached the end of an exhaustive section. We'll now go through a few more features and methods for more detailed text analysis in the next section.

More spaCy features

Most of the NLP development is token and span oriented; that is, it processes tags, dependency relations, tokens themselves, and phrases. Most of the time we eliminate small words and words without much meaning; we process URLs differently, and so on. What we do sometimes depends on the **token shape** (token is a short word or token looks like an URL string) or more semantical features (such as the token is an article, or the token is a conjunction). In this section, we will see these features of tokens with examples. We'll start with features related to the token shape:

```
doc = nlp("Hello, hi!")
doc[0].lower_
'hello'
```

`token.lower_` returns the token in lowercase. The return value is a Unicode string and this feature is equivalent to `token.text.lower()`.

`is_lower` and `is_upper` are similar to their Python string method counterparts, `islower()` and `isupper()`. `is_lower` returns `True` if all the characters are lowercase, while `is_upper` does the same with uppercase:

```
doc = nlp("HELLO, Hello, hello, hEllO")
doc[0].is_upper
True
doc[0].is_lower
False
doc[1].is_upper
False
doc[1].is_lower
False
```

`is_alpha` returns `True` if all the characters of the token are alphabetic letters. Examples of nonalphabetic characters are numbers, punctuation, and whitespace:

```
doc = nlp("Cat and Cat123")
doc[0].is_alpha
True
doc[2].is_alpha
False
```

is_ascii returns True if all the characters of token are ASCII characters.

```
doc = nlp("Hamburg and Göttingen")
doc[0].is_ascii
True
doc[2].is_ascii
False
```

is_digit returns True if all the characters of the token are numbers:

```
doc = nlp("Cat Cat123 123")
doc[0].is_digit
False
doc[1].is_digit
False
doc[2].is_digit
True
```

is_punct returns True if the token is a punctuation mark:

```
doc = nlp("You, him and Sally")
doc[1]
,
doc[1].is_punct
True
```

is_left_punct and is_right_punct return True if the token is a left punctuation mark or right punctuation mark, respectively. A right punctuation mark can be any mark that closes a left punctuation mark, such as right brackets, > or ». Left punctuation marks are similar, with the left brackets < and « as some examples:

```
doc = nlp("( [ He said yes. ] )")
doc[0]
(
doc[0].is_left_punct
True
doc[1]
[
doc[1].is_left_punct
True
doc[-1]
```

```
)
doc[-1].is_right_punct
True
doc[-2]
]
doc[-2].is_right_punct
True
```

is_space returns True if the token is only whitespace characters:

```
doc = nlp(" ")
doc[0]
len(doc[0])
1
doc[0].is_space
True
doc = nlp("  ")
doc[0]

len(doc[0])
2
doc[0].is_space
True
```

is_bracket returns True for bracket characters:

```
doc = nlp("( You said [1] and {2} is not applicable.)")
doc[0].is_bracket, doc[-1].is_bracket
(True, True)
doc[3].is_bracket, doc[5].is_bracket
(True, True)
doc[7].is_bracket, doc[9].is_bracket
(True, True)
```

`is_quote` returns `True` for quotation marks:

```
doc = nlp("( You said '1\" is not applicable.)")
doc[3]
'
doc[3].is_quote
True
doc[5]
"
doc[5].is_quote
True
```

`is_currency` returns `True` for currency symbols such as $ and € (this method was implemented by myself):

```
doc = nlp("I paid 12$ for the tshirt.")
doc[3]
$
doc[3].is_currency
True
```

`like_url`, `like_num`, and `like_email` are methods about the token shape and return `True` if the token looks like a URL, a number, or an email, respectively. These methods are very handy when we want to process social media text and scraped web pages:

```
doc = nlp("I emailed you at least 100 times")
doc[-2]
100
doc[-2].like_num
True
doc = nlp("I emailed you at least hundred times")
doc[-2]
hundred
doc[-2].like_num
True doc = nlp("My email is duygu@packt.com and you can visit
me under https://duygua.github.io any time you want.")
doc[3]
duygu@packt.com
```

```
doc[3].like_email
True
doc[10]
https://duygua.github.io/
doc[10].like_url
True
```

`token.shape_` is an unusual feature – there is nothing similar in other NLP libraries. It returns a string that shows a token's orthographic features. Numbers are replaced with d, uppercase letters are replaced with X, and lowercase letters are replaced with x. You can use the result string as a feature in your machine learning algorithms, and token shapes can be correlated to text sentiment:

```
doc = nlp("Girl called Kathy has a nickname Cat123.")
for token in doc:
    print(token.text, token.shape_)
Girl Xxxx
called xxxx
Kathy Xxxxx
has xxx
a x
nickname xxxx
Cat123 Xxxddd
. .
```

`is_oov` and `is_stop` are semantic features, as opposed to the preceding shape features. `is_oov` returns True if the token is **Out Of Vocabulary (OOV)**, that is, not in the Doc object's vocabulary. OOV words are unknown words to the language model, and thus also to the processing pipeline components:

```
doc = nlp("I visited Jenny at Mynks Resort")
for token in doc:
    print(token, token.is_oov)
I False
visited False
Jenny False
at False
Mynks True
Resort False
```

is_stop is a feature that is frequently used by machine learning algorithms. Often, we filter words that do not carry much meaning, such as *the*, *a*, *an*, *and*, *just*, *with*, and so on. Such words are called stop words. Each language has their own stop word list, and you can access English stop words here https://github.com/explosion/spaCy/blob/master/spacy/lang/en/stop_words.py:

```
doc = nlpI just want to inform you that I was with the
principle.")
for token in doc:
    print(token, token.is_stop)
I True
just True
want False
to True
inform False
you True
that True
I True
was True
with True
the True
principle False
. False
```

We have exhausted the list of spaCy's syntactic, semantic, and orthographic features. Unsurprisingly, many methods focused on the Token object as a token is the syntactic unit of a text.

Summary

We have now reached the end of an exhaustive chapter of spaCy core operations and the basic features of spaCy. This chapter gave you a comprehensive picture of spaCy library classes and methods. We made a deep dive into language processing pipelining and learned about pipeline components. We also covered a basic yet important syntactic task: tokenization. We continued with the linguistic concept of lemmatization and you learned a real-world application of a spaCy feature. We explored spaCy container classes in detail and finalized the chapter with precise and useful spaCy features. At this point, you have a good grasp of spaCy language pipelining and you are confident about accomplishing bigger tasks.

In the next chapter, we will dive into spaCy's full linguistic power. You'll discover linguistic features including spaCy's most used features: the POS tagger, dependency parser, named entities, and entity linking.

Section 2: spaCy Features

In this section, spaCy will be unveiled and examined in depth by looking at the most powerful and frequently used features. We will uncover linguistic features from syntactic to semantic, provide practical recipes with pattern matching, and advance into the semantics world with word vectors. We will also discuss statistical information extraction methods in detail. We will cover a detailed case study that shows how to combine all spaCy features to create a real-world NLP pipeline.

This section comprises the following chapters:

- *Chapter 3, Linguistic Features*
- *Chapter 4, Rule-Based Matching*
- *Chapter 5, Working with Word Vectors and Semantic Similarity*
- *Chapter 6, Putting Everything Together – Semantic Parsing with spaCy*

3
Linguistic Features

This chapter is a deep dive into the full power of spaCy. You will discover the linguistic features, including spaCy's most commonly used features such as the **part-of-speech (POS) tagger**, the **dependency parser**, the **named entity recognizer**, and **merging/splitting** features.

First, you'll learn the POS tag concept, how the spaCy POS tagger functions, and how to place POS tags into your **natural-language understanding** (NLU) applications. Next, you'll learn a structured way to represent the sentence syntax through the dependency parser. You'll learn about the dependency labels of spaCy and how to interpret the spaCy dependency labeler results with revealing examples. Then, you'll learn a very important NLU concept that lies at the heart of many **natural language processing** (NLP) applications—**named entity recognition** (NER). We'll go over examples of how to extract information from the text using NER. Finally, you'll learn how to merge/split the entities you extracted.

In this chapter, we're going to cover the following main topics:

- What is POS tagging?
- Introduction to dependency parsing
- Introducing NER
- Merging and splitting tokens

Technical requirements

The chapter code can be found at the book's GitHub repository: `https://github.com/PacktPublishing/Mastering-spaCy/tree/main/Chapter03`

What is POS tagging?

We saw the terms *POS tag* and *POS tagging* briefly in the previous chapter, while discussing the spaCy `Token` class features. As is obvious from the name, they refer to the process of tagging tokens with POS tags. One question remains here: *What is a POS tag?* In this section, we'll discover in detail the concept of POS and how to make the most of it in our NLP applications.

The **POS tagging** acronym expands as **part-of-speech tagging**. A **part of speech** is a syntactic category in which every word falls into a category according to its function in a sentence. For example, English has nine main categories: verb, noun, pronoun, determiner, adjective, adverb, preposition, conjunction, and interjection. We can describe the functions of each category as follows:

- **Verb**: Expresses an action or a state of being
- **Noun**: Identifies a person, a place, or a thing, or names a particular of one of these (a proper noun)
- **Pronoun**: Can replace a noun or noun phrase
- **Determiner**: Is placed in front of a noun to express a quantity or clarify what the noun refers to—briefly, a noun introducer
- **Adjective**: Modifies a noun or a pronoun
- **Adverb**: Modifies a verb, an adjective, or another adverb
- **Preposition**: Connects a noun/pronoun to other parts of the sentence
- **Conjunction**: Glues words, clauses, and sentences together
- **Interjection**: Expresses emotion in a sudden and exclamatory way

This core set of categories, without any language-specific morphological or syntactic features, are called **universal tags**. spaCy captures universal tags via the `pos_` feature and describes them with examples, as follows:

POS	DESCRIPTION	EXAMPLES
ADJ	adjective	big, old, green, incomprehensible, first
ADP	adposition	in, to, during
ADV	adverb	very, tomorrow, down, where, there
AUX	auxillary	is, has (done), will (do), should (do)
CONJ	conjunction	and, or, but
CCONJ	coordinating conjuction	and, or, but

Figure 3.1 – spaCy universal tags explained with examples

Throughout the book, we are providing examples with the English language and, in this section, we'll therefore focus on English. Different languages offer different tagsets, and spaCy supports different tagsets via tag_map.py under each language submodule. For example, the current English tagset lies under lang/en/tag_map.py and the German tagset lies under lang/de/tag_map.py. Also, the same language can support different tagsets; for this reason, spaCy and other NLP libraries always *specify* which tagset they use. The spaCy English POS tagger uses the Ontonotes 5 tagset, and the German POS tagger uses the TIGER Treebank tagset.

Each supported language of spaCy admits its own fine-grained tagset and tagging scheme, a specific tagging scheme that usually covers morphological features, tenses and aspects of verbs, number of nouns (singular/plural), person and number information of pronouns (first-, second-, third-person singular/plural), pronoun type (personal, demonstrative, interrogative), adjective type (comparative or superlative), and so on.

spaCy supports fine-grained POS tags to answer language-specific needs, and the `tag_` feature corresponds to the fine-grained tags. The following screenshot shows us a part of these fine-grained POS tags and their mappings to more universal POS tags for English:

CCONJ	coordinating conjuction	and, or, but
DET	determiner	a, an, the
INTJ	interjection	psst, ouch, bravo, hello
NOUN	noun	girl, cast, tree, air, beauty
NUM	numeral	1, 2017, one, seventy-seven, IV, MMXIV
PART	particle	's, not.
PRON	pronoun	I, you, he, she, myself, themselves, somebody
PROPN	proper noun	Mary, John, London, NATO, HBO

Figure 3.2 – Fine-grained English tags and universal tag mappings

Don't worry if you haven't worked with POS tags before, as you'll become familiar by practicing with the help of our examples. We'll always include explanations of the tags that we use. You can also call `spacy.explain()` on the tags. We usually call `spacy.explain()` in two ways, either directly on the tag name string or with `token.tag_`, as illustrated in the following code snippet:

```
spacy.explain("NNS)
'noun, plural'
doc = nlp("I saw flowers.")
token = doc[2]
token.text, token.tag_, spacy.explain(token.tag_)
('flowers', 'NNS', 'noun, plural')
```

If you want to know more about POS, you can read more about it at two excellent resources: *Part of Speech* at http://partofspeech.org/, and the *Eight Parts of Speech* at http://www.butte.edu/departments/cas/tipsheets/grammar/parts_of_speech.html.

As you can see, POS tagging offers a very basic syntactic understanding of the sentence. POS tags are used in NLU extensively; we frequently want to find the verbs and the nouns in a sentence and better disambiguate some words for their meanings (more on this subject soon).

Each word is tagged by a POS tag depending on its *context*—the other surrounding words and their POS tags. POS taggers are sequential statistical models, which means *that a tag of a word depends on the word-neighbor tokens, their tags, and the word itself.* POS tagging has always been done in different forms. **Sequence-to-sequence learning (Seq2seq)** started with **Hidden Markov Models (HMMs)** in the early days and evolved to neural network models—typically, **long short-term memory (LSTM)** variations (spaCy also uses an LSTM variation). You can witness the evolution of state-of-art POS tagging on the ACL website (https://aclweb.org/aclwiki/POS_Tagging_(State_of_the_art)).

It's time for some code now. Again, spaCy offers universal POS tags via the token.pos (int) and token.pos_ (unicode) features. The fine-grained POS tags are available via the token.tag (int) and token.tag_ (unicode) features. Let's learn more about tags that you'll come across most, through some examples. The following example includes examples of noun, proper noun, pronoun, and verb tags:

```python
import spacy
nlp = spacy.load("en_core_web_md")
doc = nlp("Alicia and me went to the school by bus.")
for token in doc:
    token.text, token.pos_, token.tag_, \
    spacy.explain(token.pos_), spacy.explain(token.tag_)
...
('Alicia', 'PROPN', 'NNP', 'proper noun', 'noun, proper
singular')
('and', 'CCONJ', 'CC', 'coordinating conjunction',
'conjunction, coordinating')
('me', 'PRON', 'PRP', 'pronoun', 'pronoun, personal')
('went', 'VERB', 'VBD', 'verb', 'verb, past tense')
('to', 'ADP', 'IN', 'adposition', 'conjunction, subordinating
or preposition')
```

```
('school', 'NOUN', 'NN', 'noun', 'noun, singular or mass')
('with', 'ADP', 'IN', 'adposition', 'conjunction, subordinating
or preposition')
('bus', 'NOUN', 'NN', 'noun', 'noun, singular or mass')
('.', 'PUNCT', '.', 'punctuation', 'punctuation mark, sentence
closer')
```

We iterated over the tokens and printed the tokens' text, universal tag, and fine-grained tag, together with the explanations, which are outlined here:

- `Alicia` is a proper noun, as expected, and `NNP` is a tag for proper nouns.
- `me` is a pronoun and `bus` is a noun. `NN` is a tag for singular nouns and `PRP` is a personal pronoun tag.
- Verb tags start with `V`. Here, `VBD` is a tag for *went*, which is a past-tense verb.

Now, consider the following sentence:

```
doc = nlp("My friend will fly to New York fast and she is
staying there for 3 days.")
for token in doc:
    token.text, token.pos_, token.tag_, \
    spacy.explain(token.pos_), spacy.explain(token.tag_)
...
('My', 'DET', 'PRP$', 'determiner', 'pronoun, possessive')
('friend', 'NOUN', 'NN', 'noun', 'noun, singular or mass')
('will', 'VERB', 'MD', 'verb', 'verb, modal auxiliary')
('fly', 'VERB', 'VB', 'verb', 'verb, base form')
('to', 'ADP', 'IN', 'adposition', 'conjunction, subordinating
or preposition')
('New', 'PROPN', 'NNP', 'proper noun', 'noun, proper singular')
('York', 'PROPN', 'NNP', 'proper noun', 'noun, proper
singular')
('fast', 'ADV', 'RB', 'adverb', 'adverb')
('and', 'CCONJ', 'CC', 'coordinating conjunction',
'conjunction, coordinating')
('she', 'PRON', 'PRP', 'pronoun', 'pronoun, personal')
('is', 'AUX', 'VBZ', 'auxiliary', 'verb, 3rd person singular
present')
```

```
('staying', 'VERB', 'VBG', 'verb', 'verb, gerund or present
participle')
('there', 'ADV', 'RB', 'adverb', 'adverb')
('for', 'ADP', 'IN', 'adposition', 'conjunction, subordinating
or preposition')
('3', 'NUM', 'CD', 'numeral', 'cardinal number')
('days', 'NOUN', 'NNS', 'noun', 'noun, plural')
('.', 'PUNCT', '.', 'punctuation', 'punctuation mark, sentence
closer')
```

Let's start with the verbs. As we pointed out in the first example, verb tags start with V. Here, there are three verbs, as follows:

- fly: a base form
- staying: an -*ing* form
- is: an auxiliary verb

The corresponding tags are VB, VBG, and VBZ.

Another detail is both New and York are tagged as proper nouns. If a proper noun consists of multiple tokens, then all the tokens admit the tag NNP. My is a possessive pronoun and is tagged as PRP$, in contrast to the preceding personal pronoun me and its tag PRP.

Let's continue with a word that can be a verb or noun, depending on the context: ship. In the following sentence, ship is used as a verb:

```
doc = nlp("I will ship the package tomorrow.")
for token in doc:
    token.text, token.tag_, spacy.explain(token.tag_)
...
('I', 'PRP', 'pronoun, personal')
('will', 'MD', 'verb, modal auxiliary')
('ship', 'VB', 'verb, base form')
('the', 'DT', 'determiner')
('package', 'NN', 'noun, singular or mass')
('tomorrow', 'NN', 'noun, singular or mass')
('.', '.', 'punctuation mark, sentence closer')
```

Here, `ship` is tagged as a verb, as we expected. Our next sentence also contains the word `ship`, but as a noun. Now, can the spaCy tagger tag it correctly? Have a look at the following code snippet to find out:

```
doc = nlp("I saw a red ship.")
for token in doc:
...     token.text, token.tag_, spacy.explain(token.tag_)
...
('I', 'PRP', 'pronoun, personal')
('saw', 'VBD', 'verb, past tense')
('a', 'DT', 'determiner')
('red', 'JJ', 'adjective')
('ship', 'NN', 'noun, singular or mass')
('.', '.', 'punctuation mark, sentence closer')
```

Et voilà! This time, the word `ship` is now tagged as a noun, as we wanted to see. The tagger looked at the surrounding words; here, `ship` is used with a determiner and an adjective, and spaCy deduced that it should be a noun.

How about this tricky sentence:

```
doc = nlp("My cat will fish for a fish tomorrow in a fishy way.")
for token in doc:
    token.text, token.pos_, token.tag_, \
    spacy.explain(token.pos_), spacy.explain(token.tag_)
...
('My', 'DET', 'PRP$', 'determiner', 'pronoun, possessive')
('cat', 'NOUN', 'NN', 'noun', 'noun, singular or mass')
('will', 'VERB', 'MD', 'verb', 'verb, modal auxiliary')
('fish', 'VERB', 'VB', 'verb', 'verb, base form')
('for', 'ADP', 'IN', 'adposition', 'conjunction, subordinating
or preposition')
('a', 'DET', 'DT', 'determiner', 'determiner')
('fish', 'NOUN', 'NN', 'noun', 'noun, singular or mass')
('tomorrow', 'NOUN', 'NN', 'noun', 'noun, singular or mass')
('in', 'ADP', 'IN', 'adposition', 'conjunction, subordinating
or preposition')
('a', 'DET', 'DT', 'determiner', 'determiner')
```

```
('fishy', 'ADJ', 'JJ', 'adjective', 'adjective')
('way', 'NOUN', 'NN', 'noun', 'noun, singular or mass')
('.', 'PUNCT', '.', 'punctuation', 'punctuation mark, sentence
closer')
```

We wanted to fool the tagger with the different usages of the word fish, but the tagger is intelligent enough to distinguish the verb fish, the noun fish, and the adjective fishy. Here's how it did it:

- Firstly, fish comes right after the modal verb will, so the tagger recognized it as a verb.

- Secondly, fish serves as the object of the sentence and is qualified by a determiner; the tag is most probably a noun.

- Finally, fishy ends in y and comes before a noun in the sentence, so it's clearly an adjective.

The spaCy tagger made a very smooth job here of predicting a tricky sentence. After examples of very accurate tagging, only one question is left in our minds: *Why do we need the POS tags?*

What is the importance of POS tags in NLU, and why do we need to distinguish the class of the words anyway? The answer is simple: many applications need to know the word type for better accuracy. Consider machine translation systems for an example of this: the words for fish (V) and fish (N) correspond to different words in Spanish, as illustrated in the following code snippet:

```
I will fish/VB tomorrow.    ->    Pescaré/V mañana.
I eat fish/NN.   -> Como pescado/N.
```

Syntactic information can be used in many NLU tasks, and playing some POS tricks can help your NLU code a lot. Let's continue with a traditional problem: **word-sense disambiguation (WSD)**, and how to tackle it with the help of the spaCy tagger.

WSD

WSD is a classical NLU problem of deciding in which *sense* a particular word is used in a sentence. A word can have many senses—for instance, consider the word *bass*. Here are some senses we can think of:

- Bass—seabass, fish (**noun** (N))

- Bass—lowest male voice (N)

- Bass—male singer with lowest voice range (N)

Determining the sense of the word can be crucial in search engines, machine translation, and question-answering systems. For the preceding example, *bass*, a POS tagger is unfortunately not much of help as the tagger labels all senses with a noun tag. We need more than a POS tagger. How about the word *beat*? Let's have a look at this here:

- Beat—to strike violently (**verb** (V))

- Beat—to defeat someone else in a game or a competition (V)

- Beat—rhythm in music or poetry (N)

- Beat—bird wing movement (N)

- Beat—completely exhausted (**adjective** (ADJ))

Here, POS tagging can help a lot indeed. The ADJ tag determines the word sense definitely; if the word *beat* is tagged as ADJ, it identifies the sense *completely exhausted*. This is not true for the V and N tags here; if the word *beat* is labeled with a V tag, its sense can be *to strike violently* or *to defeat someone else*. WSD is an open problem, and many complicated statistical models are proposed. However, if you need a quick prototype, you can tackle this problem in some cases (such as in the preceding example) with the help of the spaCy tagger.

Verb tense and aspect in NLU applications

In the previous chapter, we used the example of the travel agency application where we got the base forms (which are freed from verb tense and aspect) of the verbs by using **lemmatization**. In this subsection, we'll focus on how to use the verb tense and aspect information that we lost during the lemmatization process.

Verb tense and **aspect** are maybe the most interesting information that verbs provide us, telling us when the action happened in time and if the action of the verb is finished or ongoing. Tense and aspect together indicate a verb's reference to the current time. English has three basic tenses: past, present, and future. A tense is accompanied by either simple, progressive/continuous, or perfect aspects. For instance, in the sentence *I'm eating*, the action *eat* happens in the present and is ongoing, hence we describe this verb as *present progressive/continuous*.

So far, so good. So, how do we use this information in our travel agency NLU, then? Consider the following customer sentences that can be directed to our NLU application:

```
I flew to Rome.
I have flown to Rome.
I'm flying to Rome.
I need to fly to Rome.
I will fly to Rome.
```

In all the sentences, the action is *to fly*: however, only some sentences state intent to make a ticket booking. Let's imagine these sentences with a surrounding context, as follows:

```
I flew to Rome 3 days ago. I still didn't get the bill, please
send it ASAP.
I have flown to Rome this morning and forgot my laptop on the
airplane. Can you please connect me to lost and found?
I'm flying to Rome next week. Can you check flight
availability?
I need to fly to Rome. Can you check flights on next Tuesday?
I will fly to Rome next week. Can you check the flights?
```

At a quick glance, past and perfect forms of the verb *fly* don't indicate a booking intent at all. Rather, they direct to either a customer complaint or customer service issues. The infinitive and present progressive forms, on the other hand, point to booking intent. Let's tag and lemmatize the verbs with the following code segment:

```
sent1 = "I flew to Rome".
sent2 = "I'm flying to Rome."
sent3 = "I will fly to Rome."
doc1 = nlp(sent1)
doc2 = nlp(sent2)
doc3 = nlp(sent3)
for doc in [doc1, doc2, doc3]
```

```
    print([(w.text, w.lemma_) for w in doc if w.tag_ == 'VBG'
or w.tag_ == 'VB'])
...

[]
[('flying', 'fly')]
[('fly', 'fly')]
```

We iterated three doc objects one by one, and for each sentence we checked if the fine-grained tag of the token is VBG (a verb in present progressive form) or VB (a verb in base/infinitive form). Basically, we filtered out the present progressive and infinitive verbs. You can think of this process as a semantic representation of the verb in the form of (word form, lemma, tag) as illustrated in the following code snippet:

```
flying: (fly, VBG)
```

We have covered one semantic and one morphological task—WSD and tense/aspect of verbs. We'll continue with a tricky subject: how to make the best of some special tags—namely, number, symbol, and punctuation tags.

Understanding number, symbol, and punctuation tags

If you look at the English POS, you will notice the NUM, SYM, and PUNCT tags. These are the tags for numbers, symbols, and punctuation, respectively. These categories are divided into fine-grained categories: $, SYM, ' ', -LRB-, and -RRB-. These are shown in the following screenshot:

TAG	POS	MORPHOLOGY	DESCRIPTION
$	SYM		symbol, currency
' '	PUNCT	PunctType=quot PunctSide=ini	opening quotation mark
' '	PUNCT	PunctType=quot PunctSide=fin	closing quotation mark
,	PUNCT	PunctType=comm	punctuation mark, comma
-LRB-	PUNCT	PunctType=brck PunctSide=ini	left round bracket
-RRB-	PUNCT	PunctType=brck PunctSide=fin	right round bracket
,	PUNCT	PunctType=peri	punctuation mark, sentence closer
:	PUNCT		punctuation mark, colon or ellipsis

Figure 3.3 – spaCy punctuation tags, general and fine-grained

Let's tag some example sentences that contain numbers and symbols, as follows:

```
doc = nlp("He earned $5.5 million in 2020 and paid %35 tax.")
for token in doc:
    token.text, token.tag_, spacy.explain(token.tag_)
...
('He', 'PRP', 'pronoun, personal')
('earned', 'VBD', 'verb, past tense')
('$', '$', 'symbol, currency')
('5.5', 'CD', 'cardinal number')
('million', 'CD', 'cardinal number')
('in', 'IN', 'conjunction, subordinating or preposition')
('2020', 'CD', 'cardinal number')
('and', 'CC', 'conjunction, coordinating')
('paid', 'VBD', 'verb, past tense')
('35', 'CD', 'cardinal number')
('percent', 'NN', 'noun, singular or mass')
('tax', 'NN', 'noun, singular or mass')
('.', '.', 'punctuation mark, sentence closer')
```

We again iterated over the tokens and printed the fine-grained tags. The tagger was able to distinguish symbols, punctuation marks, and numbers. Even the word `million` is recognized as a number too!

Now, what to do with symbol tags? Currency symbols and numbers offer a way to systematically extract descriptions of money and are very handy in financial text such as financial reports. We'll see how to extract money entities in *Chapter 4, Rule-Based Matching*.

That's it—you made it to the end of this exhaustive section! There's a lot to unpack and digest, but we assure you that you made a great investment for your industrial NLP work. We'll now continue with another syntactic concept—dependency parsing.

Introduction to dependency parsing

If you are already familiar with spaCy, you must have come across the spaCy dependency parser. Though many developers see *dependency parser* on the spaCy documentation, they're shy about using it or don't know how to use this feature to the fullest. In this part, you'll explore a systematic way of representing a sentence syntactically. Let's start with what dependency parsing actually is.

What is dependency parsing?

In the previous section, we focused on POS tags—syntactic categories of words. Though POS tags provide information about neighbor words' tags as well, they do not give away any relations between words that are not neighbors in the given sentence.

In this section, we'll focus on dependency parsing—a more structured way of exploring the sentence syntax. As the name suggests, **dependency parsing** is related to analyzing sentence structures via dependencies between the tokens. A **dependency parser** tags syntactic relations between tokens of the sentence and connects syntactically related pairs of tokens. A **dependency** or a **dependency relation** is a *directed link* between two tokens.

The result of the dependency parsing is always a **tree**, as illustrated in the following screenshot:

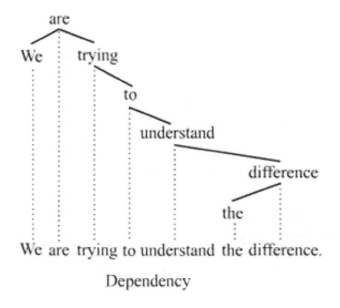

Figure 3.4 – An example of a dependency tree (taken from Wikipedia)

If you're not familiar with a tree data structure, you can learn more about it at this excellent Computer Science resource:

```
https://www.cs.cmu.edu/~clo/www/CMU/DataStructures/Lessons/
lesson4_1.htm
```

Dependency relations

What is the use of **dependency relations**, then? Quite a number of statistical methods in NLP revolve around vector representations of words and treat a sentence as a sequence of words. As you can see in *Figure 3.4*, a sentence is more than a sequence of tokens—it has a structure; every word in a sentence has a well-defined role, such as verb, subject, object, and so on; hence, sentences definitely have a structure. This structure is used extensively in chatbots, question answering, and machine translation.

The most useful application that first comes to mind is determining the sentence object and subject. Again, let's go back to our travel agency application. Imagine a customer is complaining about the service. Compare the two sentences, I forwarded you the email and You forwarded me the email; if we eliminate the stopwords I, you, me, and the, this is what remains:

```
I forwarded you the email. -> forwarded email
You forwarded me the email. -> forwarded email
```

Though the remaining parts of the sentences are identical, sentences have very different meanings and require different answers. In the first sentence, the sentence subject is I (then, the answer most probably will start with you) and the second sentence's subject is you (which will end up in an I answer).

Obviously, the dependency parser helps us to go deeper into the sentence syntax and semantics. Let's explore more, starting from the dependency relations.

Syntactic relations

spaCy assigns each token a dependency label, just as with other linguistic features such as a lemma or a POS tag. spaCy shows dependency relations with *directed arcs*. The following screenshot shows an example of a dependency relation between a noun and the adjective that qualifies the noun:

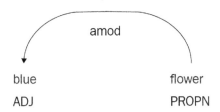

Figure 3.5 – Dependency relation between a noun and its adjective

A dependency label describes the type of syntactic relation between two tokens as follows: one of the tokens is the **syntactic parent** (called the **HEAD**) and the other is its **dependent** (called the **CHILD**). In the preceding example, `flower` is the head and `blue` is its dependent/child.

The dependency label is assigned to the child. Token objects have `dep` (`int`) and `dep_` (`unicode`) properties that hold the dependency label, as illustrated in the following code snippet:

```
doc = nlp("blue flower")
for token in doc:
    token.text, token.dep_
...
('blue', 'amod')
('flower', 'ROOT')
```

In this example, we iterated over the tokens and printed their text and dependency label. Let's go over what happened bit by bit, as follows:

- `blue` admitted the `amod` label. `amod` is the dependency label for an adjective-noun relation. For more examples of the `amod` relation, please refer to *Figure 3.7*.

- `flower` is the `ROOT`. `ROOT` is a special label in the dependency tree; it is assigned to the main verb of a sentence. If we're processing a phrase (not a full sentence), the `ROOT` label is assigned to the root of the phrase, which is the head noun of the phrase. In the `blue flower` phrase, the head noun, `flower`, is the root of the phrase.

- Each sentence/phrase has exactly one root, and it's the root of the parse tree (remember, the dependency parsing result is a tree).

- Tree nodes can have more than one child, but each node can only have one parent (due to tree restrictions, and trees containing no cycles). In other words, every token has exactly one head, but a parent can have several children. This is the reason why the dependency label is assigned to the dependent node.

Here is a full list of spaCy's English dependency labels:

LABEL	DESCRIPTION
acl	clausal modifier of noun (adjectival clause)
acomp	adjectival complement
advcl	adverbial clause modifier
advmod	adverbial modifier
agent	agent
amod	adjectival modifier
appos	appositional modifier
attr	attribute
aux	auxiliary
auxpass	auxiliary (passive)
case	case marking
cc	coordinating conjunction
ccomp	clausal complement
compound	compound
conj	conjunct
cop	copula
csubj	clausal subject
csubjpass	clausal subject (passive)
dative	dative

Figure 3.6 – List of spaCy English dependency labels

That's a long list! No worries—you don't need to memorize every list item. Let's first see a list of the most common and useful labels, then we'll see how exactly they link tokens to each other. Here's the list first:

- `amod`: Adjectival modifier
- `aux`: Auxiliary
- `compound`: Compound
- `dative`: Dative object
- `det`: Determiner
- `dobj`: Direct object
- `nsubj`: Nominal subject
- `nsubjpass`: Nominal subject, passive
- `nummod`: Numeric modifier
- `poss`: Possessive modifier
- `root`: The root

Let's see examples of how the aforementioned labels are used and what relation they express. `amod` is adjectival modifier. As understood from the name, this relation modifies the noun (or pronoun). In the following screenshot, we see **white** modifies **sheep**:

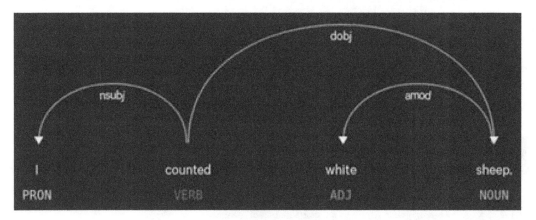

Figure 3.7 – amod relation

aux is what you might guess: it's the dependency relation between an auxiliary verb and its main verb; the dependent is an auxiliary verb, and the head is the main verb. In the following screenshot, we see that **has** is the auxiliary verb of the main verb **gone**:

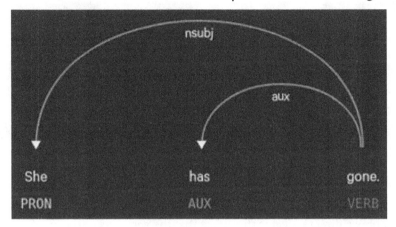

Figure 3.8 – aux relation

compound is used for the noun compounds; the second noun is modified by the first noun. In the following screenshot, **phone book** is a noun compound and the **phone** noun modifies the **book** noun:

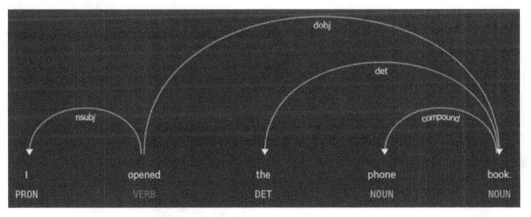

Figure 3.9 – Compound relation between phone and book

The det relation links a determiner (the dependent) to the noun it qualifies (its head). In the following screenshot, **the** is the determiner of the noun **girl** in this sentence:

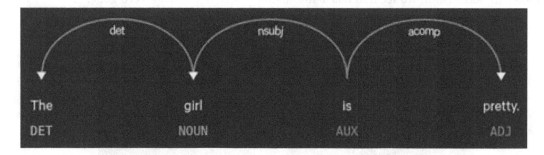

Figure 3.10 – det relation

Next, we look into two object relations, dative and dobj. The dobj relation is between the verb and its direct object. A sentence can have more than one object (such as in the following example); a direct object is the object that the verb acts upon, and the others are called indirect objects.

A direct object is generally marked with **accusative case**. A dative relation points to a dative object, which receives an indirect action from the verb. In the sentence shown in the following screenshot, the indirect object is **me** and the direct object is **book**:

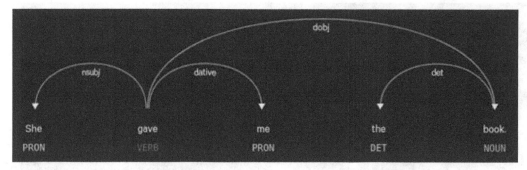

Figure 3.11 – The direct and indirect objects of the sentence

nsubj and nsubjposs are two relations that are related to the nominal sentence subject. The subject of the sentence is the one that committed the action. A passive subject is still the subject, but we mark it with nsubjposs. In the following screenshot, **Mary** is the nominal subject of the first sentence:

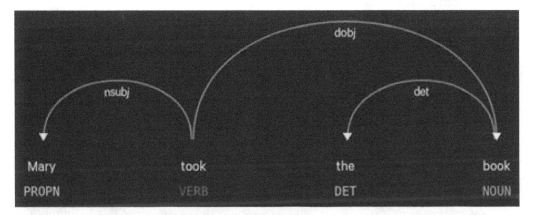

Figure 3.12 – nsubj relation

you is the passive nominal subject of the sentence in the following screenshot:

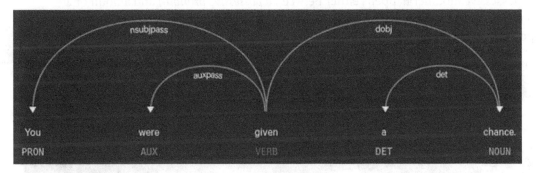

Figure 3.13 – nsubjpass relation

We have now covered sentence subject and object relations. Now, we'll discover two modifier relations; one is the nummod **numeric modifier** and the other is the poss **possessive modifier**. A numeric modifier modifies the meaning of the head noun by a number/quantity. In the sentence shown in the following screenshot, nummod is easy to spot; it's between **3** and **books**:

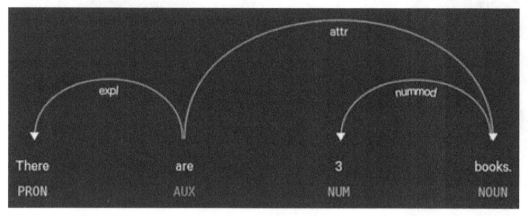

Figure 3.14 – nummod relation

A possessive modifier happens either between a *possessive pronoun* and a noun or a *possessive 's* and a noun. In the sentence shown in the following screenshot, **my** is a possessive marker on the noun **book**:

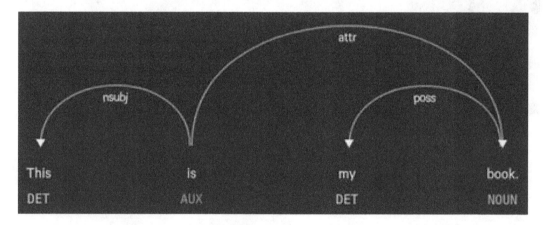

Figure 3.15 – poss relation between "my" and "book"

Last, but not least, is the **root label**, which is not a real relation but is a marker for the sentence verb. A root word has no real parent in the syntactic tree; the root is the main verb of the sentence. In the preceding sentences, **took** and **given** are the corresponding roots. The main verbs of both sentences are the auxiliary verbs **is** and **are**. Notice that the root node has no incoming arc—that is, no parent.

These are the most useful labels for our NLU purposes. You definitely don't need to memorize all the labels, as you'll become familiar as you practice in the next pages. Also, you can ask spaCy about a label any time you need, via `spacy.explain()`. The code to do this is shown in the following snippet:

```
spacy.explain("nsubj")
'nominal subject'
doc = nlp("I own a ginger cat.")
token = doc[4]
token.text, token.dep_, spacy.explain(token.dep_)
('cat', 'dobj', 'direct object')
```

Take a deep breath, since there is a lot to digest! Let's practice how we can make use of dependency labels.

Again, `token.dep_` includes the dependency label of the dependent token. The `token.head` property points to the head/parent token. Only the root token does not have a parent; spaCy points to the token itself in this case. Let's bisect the example sentence from *Figure 3.7*, as follows:

```
doc = nlp("I counted white sheep.")
for token in doc:
    token.text, token.pos_, token.dep_
...
('I', 'PRP', 'nsubj')
('counted', 'VBD', 'ROOT')
('white', 'JJ', 'amod')
('sheep', 'NNS', 'dobj')
('.', '.', 'punct')
```

We iterated over the tokens and printed the fine-grained POS tag and the dependency label. counted is the main verb of the sentence and is labeled by the label ROOT. Now, I is the subject of the sentence, and sheep is the direct object. white is an adjective and modifies the noun sheep, hence its label is amod. We go one level deeper and print the token heads this time, as follows:

```
doc = nlp("I counted white sheep.")
for token in doc:
        token.text, token.tag_, token.dep_, token.head
...
('I', 'PRP', 'nsubj', counted)
('counted', 'VBD', 'ROOT', counted)
('white', 'JJ', 'amod', sheep)
('sheep', 'NNS', 'dobj', counted)
('.', '.', 'punct', counted)
```

The visualization is as follows:

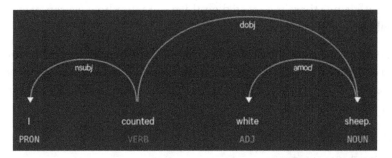

Figure 3.16 – An example parse of a simple sentence

When the token.head property is also involved, it's a good idea to follow the code and the visual at the same time. Let's go step by step in order to understand how the visual and the code match:

1. We start reading the parse tree from the root. It's the main verb: counted.

2. Next, we follow its arc on the left toward the pronoun I, which is the nominal subject of the sentence and is labeled by the label nsubj.

3. Now, return back to the root, counted. This time, we navigate to the right. Follow the dobj arc to reach the noun sheep. sheep is modified by the adjective white with an amod relation, hence the direct object of this sentence is white sheep.

Even such a simple, flat sentence has a dependency parse tree that's fancy to read, right? Don't rush—you'll get used to it by practicing. Let's examine the dependency tree of a longer and more complicated sentence, as follows:

```
doc = nlp("We are trying to understand the difference.")
for token in doc:
    token.text, token.tag_, token.dep_, token.head
...
('We', 'PRP', 'nsubj', trying)
('are', 'VBP', 'aux', trying)
('trying', 'VBG', 'ROOT', trying)
('to', 'TO', 'aux', understand)
('understand', 'VB', 'xcomp', trying)
('the', 'DT', 'det', difference)
('difference', 'NN', 'dobj', understand)
('.', '.', 'punct', trying)
```

Now, this time things look a bit different, as we'll see in *Figure 3.17*. We locate the main verb and the root `trying` (it has no incoming arcs). The left side of the word `trying` looks manageable, but the right side has a chain of arcs. Let's start with the left side. The pronoun we is labeled by `nsubj`, hence this is the nominal subject of the sentence. The other left arc, labeled `aux`, points to the `trying` dependent `are`, which is the auxiliary verb of the main verb `trying`.

So far, so good. Now, what is happening on the right side? `trying` is attached to the second verb `understand` via an `xcomp` relation. The `xcomp` (or open complement) relation of a verb is a clause without its own subject. Here, the `to understand the difference` clause has no subject, so it's an open complement. We follow the `dobj` arc from the second verb, `understand`, and land on the noun, `difference`, which is the direct object of the `to understand the difference` clause, and this is the result:

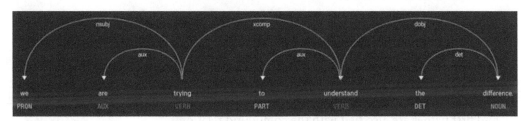

Figure 3.17 – A complicated parsing example

This was an in-depth analysis for this example sentence, which indeed does not look that complicated. Next, we process a sentence with a subsentence that owns its own nominal subject, as follows:

```
doc = nlp("Queen Katherine, who was the mother of Mary Tudor,
died at 1536.")
for token in doc:
    token.text, token.tag_, token.dep_, token.head
...
('Queen', 'NNP', 'compound', Katherine)
('Katherine', 'NNP', 'nsubj', died)
(',', ',', 'punct', Katherine)
('who', 'WP', 'nsubj', was)
('was', 'VBD', 'relcl', Katherine)
('the', 'DT', 'det', mother)
('mother', 'NN', 'attr', was)
('of', 'IN', 'prep', mother)
('Mary', 'NNP', 'compound', Tudor)
('Tudor', 'NNP', 'pobj', of)
(',', ',', 'punct', Katherine)
('died', 'VBD', 'ROOT', died)
('at', 'IN', 'prep', died)
('1536', 'CD', 'pobj', at)
```

In order to make the visuals big enough, I have split the visualization into two parts. First, let's find the root. The root lies in the right part. `died` is the main sentence of the verb and the root (again, it has no incoming arcs). The rest of the right side contains nothing tricky.

On the other hand, the left side has some interesting stuff—actually, a relative clause. Let's bisect the relative clause structure:

- We start with the proper noun `Katherine`, which is attached to `died` with a `nsubj` relation, hence the subject of the sentence.

- We see a compound arc leaving `Katherine` toward the proper noun, `Queen`. Here, `Queen` is a title, so the relationship with `Katherine` is compound. The same relationship exists between `Mary` and `Tudor` on the right side, and the last names and first names are also tied with the compound relation.

It's time to bisect the relative clause, `who was the mother of Mary Tudor,` as follows:

- First of all, it is `Katherine` who is mentioned in the relative clause, so we see a `relcl` (relative clause) arc from `Katherine` to `was` of the relative clause.

- `who` is the nominal subject of the clause and is linked to `was` via an `nsubj` relation. As you see in the following screenshot, the dependency tree is different from the previous example sentence, whose clause didn't own a nominal subject:

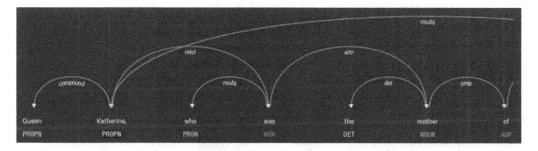

Figure 3.18 – A dependency tree with a relative clause, the left part

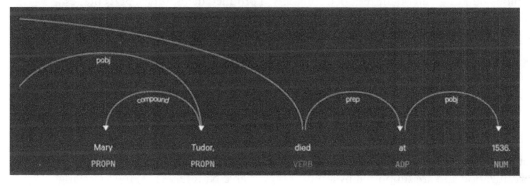

Figure 3.19 – Same sentence, the right part

It's perfectly normal if you feel that you won't be able to keep all the relations in your mind. No worries—always find the root/main verb of the sentence, then follow the arcs from the root and go deeper, just as we did previously. You can always have a look at the spaCy documentation (`https://spacy.io/api/annotation#dependency-parsing`) to see what the relation type means. Take your time until you warm up to the concept and the details.

That was exhaustive! Dear reader—as we said before, please take your time to digest and practice on example sentences. The *displaCy* online demo is a great tool, so don't be shy to try your own example sentences and see the parsing results. It's perfectly normal for you to find this section heavy. However, this section is a solid foundation for general linguistics, and also for the information extraction and pattern-matching exercises in *Chapter 4, Rule-Based Matching*. You will become even more comfortable after going through a case study in *Chapter 6, Putting Everything Together: Semantic Parsing with spaCy*. Give yourself time to digest dependency parsing with examples throughout the book.

What comes after the dependency parser? Without any doubt, you must have heard NER frequently mentioned in the NLU world. Let's look into this very important NLU concept.

Introducing NER

We opened this chapter with a tagger, and we'll see another very handy tagger—the NER tagger of spaCy. As NER's name suggests, we are interested in finding named entities.

What is a **named entity**? A named entity is a real-world object that we can refer to by a proper name or a quantity of interest. It can be a person, a place (city, country, landmark, famous building), an organization, a company, a product, dates, times, percentages, monetary amounts, a drug, or a disease name. Some examples are Alicia Keys, Paris, France, Brandenburg Gate, WHO, Google, Porsche Cayenne, and so on.

A named entity always points to a *specific* object, and that object is distinguishable via the corresponding named entity. For instance, if we tag the sentence *Paris is the capital of France*, we parse *Paris* and *France* as named entities, but not the word *capital*. The reason is that *capital* does not point to a specific object; it's a general name for many objects.

NER categorization is a bit different from POS categorization. Here, the number of categories is as high as we want. The most common categories are person, location, and organization and are supported by almost every usable NER tagger. In the following screenshot, we see the corresponding tags:

TYPE	DESCRIPTION
PER	Named person or family.
LOC	Name of politically or geographically defined location (cities, provinces, countries, international regions, bodies of water, mountains).
ORG	Named corporate, governmental, or other organizational entity.
MISC	Miscellaneous entities, e.g. events, nationalities, products or works or art.

Figure 3.20 – Most common entity types

spaCy supports a wide range of entity types. Which ones you use depends on your corpus. If you process financial text, you most probably use MONEY and PERCENTAGE more often than WORK_OF_ART.

Here is a list of the entity types supported by spaCy:

TYPE	DESCRIPTION
PERSON	People, including fictional.
NORP	Nationalities or religious or political groups.
FAC	Buildings, airports, highways, bridges, etc.
ORG	Companies, agencies, institutions, etc.
GPE	Countries, cities, states.
LOC	Non-GPE locations, mountain ranges, bodies of water.
PRODUCT	Objects, vehicles, foods, etc. (Not services.)
EVENT	Named hurricanes, battlesm wars, sports events, etc.
WORK_OF_ART	Titles of books, songs, etc.
LAW	Named documents made into laws.
LANGUAGE	Any named language.
DATE	Absolute or relative dates or periods.
TIME	Times smaller than a day.
PERCENT	Percentage, including "%".
MONEY	Monetary values, including unit.
QUANTITY	Measurements, as of weight or distance.
ORDINAL	"first", "second", etc.
CARDINAL	Numerals that do not fall under another type.

Figure 3.21 – Full list of entity types supported by spaCy

Just as with the POS tagger statistical models, NER models are also sequential models. The very first modern NER tagger model is a **conditional random field** (**CRF**). CRFs are sequence classifiers used for structured prediction problems such as labeling and parsing. If you want to learn more about the CRF implementation details, you can read more at this resource: `https://homepages.inf.ed.ac.uk/csutton/publications/crftutv2.pdf`. The current state-of-the-art NER tagging is achieved by neural network models, usually LSTM or LSTM+CRF architectures.

Named entities in a doc are available via the `doc.ents` property. `doc.ents` is a list of `Span` objects, as illustrated in the following code snippet:

```
doc = nlp("The president Donald Trump visited France.")
doc.ents
(Donald Trump, France)
type(doc.ents[1])
<class 'spacy.tokens.span.Span'>
```

spaCy also tags each token with the entity type. The type of the named entity is available via `token.ent_type` (int) and `token.ent_type_` (unicode). If the token is not a named entity, then `token.ent_type_` is just an empty string.

Just as for POS tags and dependency labels, we can call `spacy.explain()` on the tag string or on the `token.ent_type_`, as follows:

```
spacy.explain("ORG")
'Companies, agencies, institutions, etc.
doc = nlp("He worked for NASA.")
token = doc[3]
token.ent_type_, spacy.explain(token.ent_type_)
('ORG', 'Companies, agencies, institutions, etc.')
```

Let's go over some examples to see the spaCy NER tagger in action, as follows:

```
doc = nlp("Albert Einstein was born in Ulm on 1879. He studied
electronical engineering at ETH Zurich.")
doc.ents
(Albert Einstein, Ulm, 1879, ETH Zurich)
for token in doc:
    token.text, token.ent_type_, \
    spacy.explain(token.ent_type_)
...
```

```
('Albert', 'PERSON', 'People, including fictional')
('Einstein', 'PERSON', 'People, including fictional')
('was', '', None)
('born', '', None)
('in', '', None)
('Ulm', 'GPE', 'Countries, cities, states')
('on', '', None)
('1879', 'DATE', 'Absolute or relative dates or periods')
('.', '', None)
('He', '', None)
('studied', '', None)
('electronical', '', None)
('engineering', '', None)
('at', '', None)
('ETH', 'ORG', 'Companies, agencies, institutions, etc.')
('Zurich', 'ORG', 'Companies, agencies, institutions, etc.')
('.', '', None)
```

We iterated over the tokens one by one and printed the token and its entity type. If the token is not tagged as an entity, then token.ent_type_ is just an empty string, hence there is no explanation from spacy.explain(). For the tokens that are part of a NE, an appropriate tag is returned. In the preceding sentences, Albert Einstein, Ulm, 1879, and ETH Zurich are correctly tagged as PERSON, GPE, DATE, and ORG, respectively.

Let's see a longer and more complicated sentence with a non-English entity and look at how spaCy tagged it, as follows:

```
doc = nlp("Jean-Michel Basquiat was an American artist of
Haitian and Puerto Rican descent who gained fame with his
graffiti and street art work")
doc.ents
(Jean-Michel Basquiat, American, Haitian, Puerto Rican)
for ent in doc.ents:
    ent, ent.label_, spacy.explain(ent.label_)
...
(Jean-Michel Basquiat, 'PERSON', 'People, including fictional')
(American, 'NORP', 'Nationalities or religious or political
groups')
```

```
(Haitian, 'NORP', 'Nationalities or religious or political
groups')
(Puerto Rican, 'NORP', 'Nationalities or religious or political
groups')
```

Looks good! The spaCy tagger picked up a person entity with a - smoothly. Overall, the tagger works quite well for different entity types, as we saw throughout the examples.

After tagging tokens with different syntactical features, we sometimes want to merge/split entities into fewer/more tokens. In the next section, we will see how merging and splitting is done. Before that, we will see a real-world application of NER tagging.

A real-world example

NER is a popular and frequently used pipeline component of spaCy. NER is one of the key components of understanding the text topic, as named entities usually belong to a **semantic category**. For instance, *President Trump* invokes the *politics* subject in our minds, whereas *Leonardo DiCaprio* is more about *movies*. If you want to go deeper into resolving the text meaning and understanding who made what, you also need named entities.

This real-world example includes processing a *New York Times* article. Let's go ahead and download the article first by running the following code:

```
from bs4 import BeautifulSoup
import requests
import spacy
def url_text(url_string):
    res = requests.get(url_string)
    html = res.text
    soup = BeautifulSoup(html, 'html5lib')
    for script in soup(["script", "style", 'aside']):
        script.extract()
    text = soup.get_text()
    return " ".join(text.split())
ny_art = url_text("https://www.nytimes.com/2021/01/12/opinion/
trump-america-allies.html")

nlp = spacy.load("en_core_web_md")
doc = nlp(ny_art)
```

We downloaded the article **Uniform Resource Locator** (**URL**), and then we stripped the article text from the **HyperText Markup Language** (**HTML**). BeautifulSoup is a popular Python package for extracting text from HTML and **Extensible Markup Language** (**XML**) documents. Then, we created a nlp object, passed the article body to the nlp object, and created a Doc object.

Let's start our analysis of the article by the entity type count, as follows:

```
len(doc.ents)
136
```

That's a totally normal number for a news article that includes many entities. Let's go a bit further and group the entity types, as follows:

```
from collections import Counter
labels = [ent.label_ for ent in doc.ents]
Counter(labels)
Counter({'GPE': 37, 'PERSON': 30, 'NORP': 24, 'ORG': 22,
'DATE': 13, 'CARDINAL': 3, 'FAC': 2, 'LOC': 2, 'EVENT': 1,
'TIME': 1, 'WORK_OF_ART': 1})
```

The most frequent entity type is GPE, which means a country, city, or state. The second one is PERSON, whereas the third most frequent entity label is NORP, which means a nationality/religious-political group. The next ones are organization, date, and cardinal number-type entities.

Can we summarize the text by looking at the entities or understanding the text topic? To answer this question, let's start by counting the most frequent tokens that occur in the entities, as follows:

```
items = [ent.text for ent in doc.ents]
Counter(items).most_common(10)
[('America', 12), ('American', 8), ('Biden', 8), ('China',
6), ('Trump', 5), ('Capitol', 4), ('the United States', 3),
('Washington', 3), ('Europeans', 3), ('Americans', 3)]
```

Looks like a semantic group! Obviously, this article is about American politics, and possibly how America interacts with the rest of the world in politics. If we print all the entities of the article, we can see here that this guess is true:

```
print(doc.ents)
```

```
(The New York Times SectionsSEARCHSkip, indexLog inToday,
storyOpinionSupported byContinue, LaughingstockLast week's,
U.S., U.S., Ivan KrastevMr, Krastev, Jan., 2021
阅读简体中文版閱讀繁體中文版 A, Rome, Donald Tramp, Thursday, Andrew
Medichini, Associated PressDonald Trump, America, America,
Russian, Chinese, Iranian, Jan. 6, Capitol, Ukraine, Georgia,
American, American, the United States, Trump, American,
Congress, Civil War, 19th-century, German, Otto von Bismarck,
the United States of America, America, Capitol, Trump, last
hours, American, American, Washington, Washington, Capitol,
America, America, Russia, at least 10, Four years, Trump, Joe
Biden, two, American, China, Biden, America, Trump, Recep
Tayyip Erdogan, Turkey, Jair Bolsonaro, Brazil, Washington,
Russia, China, Biden, Gianpaolo Baiocchi, H. Jacob Carlson,
Social Housing Development Authority, Ezra Klein, Biden, Mark
Bittman, Biden, Gail Collins, Joe Biden, Jake Sullivan, Biden,
trans-Atlantic, China, Just a week ago, European, Sullivan,
Europe, America, China, Biden, Europeans, China, German,
Chinese, the European Union's, America, Christophe Ena, the
European Council on Foreign Relations, the weeks, American, the
day, Biden, Europeans, America, the next 10 years, China, the
United States, Germans, Trump, Americans, Congress, America,
Bill Clinton, Americans, Biden, the White House, the United
States, Americans, Europeans, the past century, America, the
days, Capitol, democratic, Europe, American, America, Ivan
Krastev, the Center for Liberal Strategies, the Institute for
Human Sciences, Vienna, Is It Tomorrow Yet?:, The New York
Times Opinion, Facebook, Twitter (@NYTopinion, Instagram,
AdvertisementContinue, IndexSite Information Navigation© 2021,
The New York Times, GTM, tfAzqo1rYDLgYhmTnSjPqw&gtm_preview)
```

We made a visualization of the whole article by pasting the text into **displaCy Named Entity Visualizer** (`https://explosion.ai/demos/displacy-ent/`). The following screenshot is taken from the demo page that captured a part of the visual:

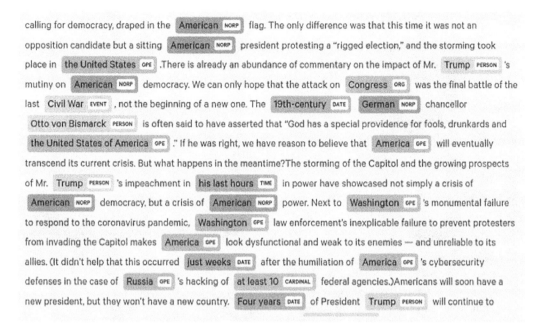

Figure 3.22 – The New York Times article's entities visualized by displaCy

spaCy's NER offers great capabilities for understanding text, as well as presenting good-looking visuals to ourselves, colleagues, and stakeholders.

Merging and splitting tokens

We extracted the name entities in the previous section, but how about if we want to unite or split multiword named entities? And what if the tokenizer performed this not so well on some exotic tokens and you want to split them by hand? In this subsection, we'll cover a very practical remedy for our multiword expressions, multiword named entities, and typos.

`doc.retokenize` is the correct tool for merging and splitting the spans. Let's see an example of retokenization by merging a multiword named entity, as follows:

```
doc = nlp("She lived in New Hampshire.")
doc.ents
(New Hampshire,)
[(token.text, token.i) for token in doc]
[('She', 0), ('lived', 1), ('in', 2), ('New', 3), ('Hampshire',
4), ('.', 5)]
len(doc)
```

```
6
with doc.retokenize() as retokenizer:
    retokenizer.merge(doc[3:5], \
    attrs={"LEMMA": "new hampshire"})
...
[(token.text, token.i) for token in doc]
[('She', 0), ('lived', 1), ('in', 2), ('New Hampshire', 3),
('.', 4)]
len(doc)
5
doc.ents
(New Hampshire,)
[(token.lemma_) for token in doc]
['-PRON-', 'live', 'in', 'new hampshire', '.']
```

This is what we did in the preceding code:

1. First, we created a doc object from the sample sentence.

2. Then, we printed its entities with doc.ents, and the result was New Hampshire, as expected.

3. In the next line, for each token, we printed token.text with token indices in the sentence (token.i).

4. Also, we examined length of the doc object by calling len on it, and the result was 6 ("." is a token too).

Now, we wanted to merge the tokens of position 3 until 5 (3 is included; 5 is not), so we did the following:

1. First, we called the retokenizer method merge(indices, attrs). attrs is a dictionary of token attributes we want to assign to the new token, such as lemma, pos, tag, ent_type, and so on.

2. In the preceding example, we set the lemma of the new token; otherwise, the lemma would be New only (the starting token's lemma of the span we want to merge).

3. Then, we printed the tokens to see if the operation worked as we wished. When we print the new tokens, we see that the new doc[3] is the New Hampshire token.

4. Also, the `doc` object is of length 5 now, so we shrank the doc one less token. `doc.` `ents` remain the same and the new token's lemma is `new hampshire` because we set it with `attrs`.

Looks good, so how about splitting a multiword token into several tokens? In this setting, either there's a typo in the text you want to fix or the custom tokenization is not satisfactory for your specific sentence.

Splitting a span is a bit more complicated than merging a span because of the following reasons:

- We are changing the dependency tree.

- We need to assign new POS tags, dependency labels, and necessary token attributes to the new tokens.

- Basically, we need to think about how to assign linguistic features to the new tokens we created.

Let's see how to deal with the new tokens with an example of how to fix a typo, as follows:

```
doc = nlp("She lived in NewHampshire")
len(doc)
5
 [(token.text, token.lemma_, token.i) for token in doc]
[('She', '-PRON-', 0), ('lived', 'live', 1), ('in', 'in', 2),
('NewHampshire', 'NewHampshire', 3), ('.', '.', 4)]
 for token in doc:
      token.text, token.pos_, token.tag_, token.dep_
...
('She', 'PRON', 'PRP', 'nsubj')
('lived', 'VERB', 'VBD', 'ROOT')
('in', 'ADP', 'IN', 'prep')
('NewHampshire', 'PROPN', 'NNP', 'pobj')
('.', 'PUNCT', '.', 'punct')
```

Here's what the dependency tree looks like before the splitting operation:

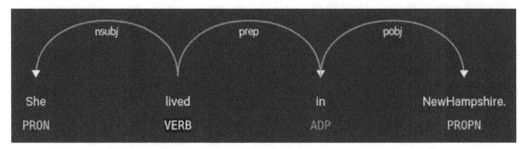

Figure 3.23 – Sample sentence's dependency tree before retokenization

Now, we will split the doc[3], NewHampshire, into two tokens: New and Hampshire. We will give fine-grained POS tags and dependency labels to the new tokens via the attrs dictionary. We will also rearrange the dependency tree by passing the new tokens' parents via the heads parameter. While arranging the heads, there are two things to consider, as outlined here:

- Firstly, if you give a relative position (such as (doc[3], 1)) in the following code segment, this means that head of doc[3] will be the +1th position token—that is, doc[4] in the new setup (please see the following visualization).

- Secondly, if you give an absolute position, it means the position in the *original* Doc object. In the following code snippet, the second item in the heads list means that the Hampshire token's head is the second token in the original Doc, which is the in token (please refer to *Figure 3.23*).

After the splitting, we printed the list of new tokens and the linguistic attributes. Also, we examined the new length of the doc object, which is 6 now. You can see the result here:

```
with doc.retokenize() as retokenizer:
    heads = [(doc[3], 1), doc[2]]
    attrs = {"TAG":["NNP", "NNP"],
             "DEP": ["compound", "pobj"]}
    retokenizer.split(doc[3], ["New", "Hampshire"],
                      heads=heads, attrs=attrs)
...
 [(token.text, token.lemma_, token.i) for token in doc]
[('She', '-PRON-', 0), ('lived', 'live', 1), ('in', 'in', 2),
('New', 'New', 3), ('Hampshire', 'Hampshire', 4), ('.', '.',
5)]
```

```
for token in doc:
    token.text, token.pos_, token.tag_, token.dep_
...
('She', 'PRON', 'PRP', 'nsubj')
('lived', 'VERB', 'VBD', 'ROOT')
('in', 'ADP', 'IN', 'prep')
('New', 'PROPN', 'NNP', 'pobj')
('Hampshire', 'PROPN', 'NNP', 'compound')
('.', 'PUNCT', '.', 'punct')
 len(doc)
6
```

Here's what the dependency tree looks like after the splitting operation (please compare this with *Figure 3.22*):

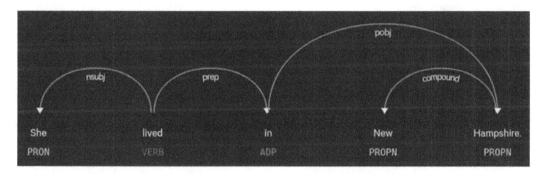

Figure 3.24 – Dependency tree after the splitting operation

You can apply merging and splitting onto any span, not only the named entity spans. The most important part here is to correctly arrange the new dependency tree and the linguistic attributes.

Summary

That was it—you made it to the end of this chapter! It was an exhaustive and long journey for sure, but we have unveiled the real linguistic power of spaCy to the fullest. This chapter gave you details of spaCy's linguistic features and how to use them.

You learned about POS tagging and applications, with many examples. You also learned about an important yet not so well-known and well-used feature of spaCy—the dependency labels. Then, we discovered a famous NLU tool and concept, NER. We saw how to do named entity extraction, again via examples. We finalized this chapter with a very handy tool for merging and splitting the spans that we calculated in the previous sections.

What's next, then? In the next chapter, we will again be discovering a spaCy feature that you'll be using every day in your NLP application code—spaCy's `Matcher` class. We don't want to give a spoiler on this beautiful subject, so let's go onto our journey together!

4
Rule-Based Matching

Rule-based information extraction is indispensable for any NLP pipeline. Certain types of entities, such as times, dates, and telephone numbers have distinct formats that can be recognized by a set of rules, without having to train statistical models.

In this chapter, you will learn how to quickly extract information from the text by matching patterns and phrases. You will use **morphological features**, **POS tags**, **regex**, and other spaCy features to form pattern objects to feed to the Matcher objects. You will continue with fine-graining statistical models with rule-based matching to lift statistical models to better accuracies.

By the end of this chapter, you will know a vital part of information extraction. You will be able to extract entities of specific formats, as well as entities specific to your domain.

In this chapter, we're going to cover the following main topics:

- Token-based matching
- PhraseMatcher
- EntityRuler
- Combining spaCy models and matchers

Token-based matching

So far, we've explored the sophisticated linguistic concepts that require statistical models and their usages with spaCy. Some NLU tasks can be solved in tricky ways without the help of any statistical model. One of those ways is **regex**, which we use to match a predefined set of patterns to our text.

A regex (a regular expression) is a sequence of characters that specifies a search pattern. A regex describes a set of strings that follows the specified pattern. A regex can include letters, digits, and characters with special meanings, such as ?, ., and *. Python's built-in library provides great support to define and match regular expressions. There's another Python 3 library called regex that aims wants to replace **re** in the future.

Readers who are actively developing NLP applications with Python have definitely come across regex code and, even better, have written regex themselves.

What does a regex look like, then? The following regex matches the following strings:

- Barack Obama
- Barack Obama
- Barack Hussein Obama

```
reg = r"Barack\s(Hussein\s)?Obama"
```

This pattern can be read as: the string `Barack` can be followed optionally by the string `Hussein` (the ? character in regex means optional, that is, 0 or 1 occurrence) and should be followed by the string `Obama`. The inter-word spaces can be a single space character, a tab, or any other whitespace character (\s matches all sorts of whitespace characters, including the newline character).

It's not very readable, even for such a short and uncomplicated pattern, is it? That is the downside of regex, it is the following:

- Difficult to read
- Difficult to debug
- Error prone with space, punctuation, and number characters

For these reasons, many software engineers don't like to work with regex in their production code. spaCy provides a very clean, readable, production-level, and maintainable alternative: the `Matcher` class. The `Matcher` class can match our predefined rules to the sequence of tokens in `Doc` and `Span` objects; moreover, the rules can refer to the token or its linguistic attributes (more on this subject later in this section).

Let's start with a basic example of how to call the `Matcher` class:

```
import spacy
from spacy.matcher import Matcher
nlp = spacy.load("en_core_web_md")
doc = nlp("Good morning, I want to reserve a ticket.")
matcher = Matcher(nlp.vocab)
pattern = [{"LOWER": "good"}, {"LOWER": "morning"},
           {"IS_PUNCT": True}]
matcher.add("morningGreeting", [pattern])
matches = matcher(doc)
for match_id, start, end in matches:
    m_span = doc[start:end]
    print(start, end, m_span.text)
...
0 3 Good morning,
```

It looks complicated, but don't be intimidated, we'll go over the lines one by one:

- We imported `spacy` in the first line; this should be familiar.

- On the second line, we imported the `Matcher` class in order to use it in the rest of the code.

- On the next lines, we created the `nlp` object as usual and created the `doc` object with our example sentence.

- Now, pay attention: a `matcher` object needs to be initialized with a `Vocabulary` object, so on line 5 we initialize our `matcher` object with the language model's vocabulary (this is the usual way to do it).

- What comes next is to define the pattern we want to match. Here, we define *pattern* as a list where every list item enclosed in a bracelet represents one token object.

You can read the pattern list in the preceding code snippet as follows:

1. A token whose lowered text is good

2. A token whose lowered text is morning

3. A token that is punctuation (that is, the `IS_PUNCT` feature is `True`)

Then, we need to introduce this pattern to the `matcher`; this is what the `matcher.add()` line does. On line 7, we introduced our pattern to the `matcher` object and named this rule `morningGreeting`. Finally, we can do the matching operation on line 8 by calling `matcher` on the `doc`. After that, we examine the result we get. A match result is a list of triplets in the form (`match id, start position, end position`). On the final line, we iterate over the result list and print the result match's start position, end position, and text.

As you might have noticed, the whitespace between `Good` and `morning` didn't matter at all. Indeed, we could have put two whitespaces in between, written down `Good morning`, and the result would be identical. Why? Because `Matcher` matches the tokens and the token attributes.

A pattern always refers to a continuous sequence of token objects, and every item in bracelets corresponds to one token object. Let's go back to the pattern in the preceding code snippet:

```
pattern = [{"LOWER": "good"}, {"LOWER": "morning"},
           {"IS_PUNCT": True}]
```

We see that the result is always a three-token match.

Can we add more than one pattern? The answer is yes. Let's see it with an example and also see an example of `match_id` as follows:

```
import spacy
from spacy.matcher import Matcher
nlp = spacy.load("en_core_web_md")
doc = nlp("Good morning, I want to reserve a ticket. I will
then say good evening!")
matcher = Matcher(nlp.vocab)
pattern1 = [{"LOWER": "good"}, {"LOWER": "morning"},
            {"IS_PUNCT": True}]
matcher.add("morningGreeting", [pattern1])
pattern2 = [{"LOWER": "good"}, {"LOWER": "evening"},
            {"IS_PUNCT": True}]
matcher.add("eveningGreeting", [pattern2])
matches = matcher(doc)
for match_id, start, end in matches:
    pattern_name = nlp.vocab_strings[match_id]
    m_span = doc[start:end]
```

```
        print(pattern_name, start, end, m_span.text)
...
morningGreeting 0 3 Good morning,
eveningGreeting 15 18 good evening!
```

This time we did things a bit differently:

- On line 8, we defined a second pattern, again matching three tokens, but this time evening instead of morning.

- On the next line, we added it to the matcher. At this point, matcher contains 2 patterns: morningGreeting and eveningGreeting.

- Again, we called the matcher on our sentence and examined the result. This time the results list has two items, Good morning, and good evening!, corresponding to two different patterns, morningGreeting and eveningGreeting.

In the preceding code example, pattern1 and pattern2 differ only by one token: evening/morning. Instead of writing two patterns, can we say evening or morning? We can do that as well. Here are the attributes that Matcher recognizes:

ATTRIBUTE	TYPE	DESCRIPTION
ORTH	unicode	The exact verbatim text of a token.
TEXT V2.1	unicode	The exact verbatim text of a token.
LOWER	unicode	The lowercase form of the token text.
LENGTH	int	The length of the token text.
IS_ALPHA , IS_ASCII , IS_DIGIT	bool	Token text consists of alphabetic characters, ASCII characters, digits.
IS_LOWER , IS_UPPER , IS_TITLE	bool	Token text is in lowercase, uppercase, titlecase.
IS_PUNCT , IS_SPACE , IS_STOP	bool	Token is punctuation, whitespace, stop word.
IS_SENT_START	bool	Token is start of sentence.
LIKE_NUM , LIKE_URL , LIKE_EMAIL	bool	Token text resembles a number, URL, email.
POS , TAG , DEP , LEMMA , SHAPE	unicode	The token's simple and extended part-of-speech tag, dependency label, lemma, shape.
ENT_TYPE	unicode	The token's entity label.
_ V2.1	dict	Properties in custom extension attributes.

Figure 4.1 – Token attributes for Matcher

Let's go over the attributes one by one with some examples. We used LOWER in the preceding examples; it means the *lowercase form of the token text*. ORTH and TEXT are similar to LOWER: they mean an exact match of the token text, including the case. Here's an example:

```
pattern = [{"TEXT": "Bill"}]
```

The preceding code will match BIll, but not bill. LENGTH is used for specifying the token length. The following code finds all tokens of length 1:

```
doc = nlp("I bought a pineapple.")
matcher = Matcher(nlp.vocabulary)
pattern = [{"LENGTH": 1}]
matcher.add("onlyShort", [pattern])
matches = matcher(doc)
for mid, start, end in matches:
    print(start, end, doc[start:end])
...
0 1 I
2 3 a
```

The next block of token attributes is IS_ALPHA, IS_ASCII, and IS_DIGIT. These features are handy for finding number tokens and *ordinary* words (which do not include any interesting characters). The following pattern matches a sequence of two tokens, a number followed by an ordinary word:

```
doc1 = nlp("I met him at 2 o'clock.")
doc2 = nlp("He brought me 2 apples.")
pattern = [{"IS_DIGIT": True},{"IS_ALPHA": True}]
matcher.add("numberAndPlainWord", [pattern])
matcher(doc1)
[]
matches = matcher(doc2)
len(matches)
1
mid, start, end = matches[0]
print(start, end, doc2[start:end])
3, 5, 2 apples
```

In the preceding code segment, 2 o'clock didn't match the pattern because o'clock contains an apostrophe, which is not an alphabetic character (alphabetic characters are digits, letters, and the underscore character). 2 apples matched because the token apples consists of letters.

IS_LOWER, IS_UPPER, and IS_TITLE are useful attributes for recognizing the token's casing. IS_UPPER is True if the token is all uppercase letters and IS_TITLE is True if the token starts with a capital letter. IS_LOWER is True if the token is all lowercase letters. Imagine we want to find emphasized words in a text; one way is to look for the tokens with all uppercase letters. The uppercase tokens usually have significant weights in sentiment analysis models.

```
doc = nlp("Take me out of your SPAM list. We never asked you
to contact me. If you write again we'll SUE!!!!")
pattern = [{"IS_UPPER": True}]
matcher.add("capitals", [pattern])
matches = matcher(doc)
for mid, start, end in matches:
    print(start, end, doc[start:end])
...
5, 6, SPAM
22, 23, SUE
```

IS_PUNCT, IS_SPACE, and IS_STOP are usually used in patterns that include some helper tokens and correspond to punctuation, space, and **stopword** tokens (stopwords are common words of a language that do not carry much information, such as *a*, *an*, and *the* in English). IS_SENT_START is another useful attribute; it matches sentence start tokens. Here's a pattern for sentences that start with *can* and the second word has a capitalized first letter:

```
doc1 = nlp("Can you swim?")
doc2 = nlp("Can Sally swim?")
pattern = [{"IS_SENT_START": True, "LOWER": "can"},
           {"IS_TITLE": True}]
matcher.add("canThenCapitalized", [pattern])
matcher(doc)
[]
matches = matcher(doc2)
len(matches)
1
```

```
 mid, start, end = matches[0]
 print(start, end, doc2[start:end])
0, 2, Can Sally
```

Here, we did a different thing: we put two attributes into one brace. In this example, the first item in `pattern` means that a token that is the first token of the sentence and whose lowered text is *can*. We can add as many attributes as we like. For instance, `{"IS_SENT_START": False, "IS_TITLE": True, "LOWER": "bill"}` is a completely valid attribute dictionary, and it describes a token that is capitalized, not the first token of sentence, and has the text `bill`. So, it is the set of `Bill` instances that does not appear as the first word of a sentence.

`LIKE_NUM`, `LIKE_URL`, and `LIKE_EMAIL` are attributes that are related to token shape again; remember, we saw them in *Chapter 3, Linguistic Features*. These attributes match tokens that look like numbers, URLs, and emails.

Though the preceding code looks short and simple, the shape attributes can be lifesavers in NLU applications. Most of the time you need nothing other than clever combinations of shape and linguistic attributes.

After seeing the shape attributes, let's see the `POS`, `TAG`, `DEP`, `LEMMA`, and `SHAPE` linguistic attributes. You saw these token attributes in the previous chapter; now we'll use them in token matching. The following code snippet spots sentences that start with an auxiliary verb:

```
 doc = nlp("Will you go there?')
 pattern = [{"IS_SENT_START": True, "TAG": "MD"}]
 matcher.add([pattern])
 matches = matcher(doc)
 len(matches)
1
 mid, start, end = matches[0]
 print(start, end, doc[start:end])
0, 1, Will
 doc2 = nlp("I might go there.")
 matcher(doc2)
[]
```

You may recall from *Chapter 3, Linguistic Features*, that MD is the tag for modal and auxiliary verbs. The preceding code snippet is a standard way of finding yes/no question sentences. In such cases, we usually look for sentences that start with a modal or an auxiliary verb.

Pro tip

Don't be afraid to work with **parts-of-speech** (**POS**) tags, they're your friends and can be lifesavers in some situations. When we want to extract the meaning of a word, we usually combine TEXT/LEMMA with POS/TAG. For instance, the word *match* is *to go together* when it's a verb or it can be a *fire starter tool* when it's a noun. In this case, we make the distinction as follows:

{"LEMMA": "match", "POS": "VERB"} and

{"LEMMA": "match", "POS": "NOUN".

Similarly, you can combine other linguistic features with token shape attributes to make sure that you extract only the pattern you mean to.

We'll see more examples of combining linguistic features with the Matcher class in the upcoming sections. Now, we'll explore more Matcher features.

Extended syntax support

Matcher allows patterns to be more expressive by allowing some operators inside the curly brackets. These operators are for extended comparison and look similar to Python's in, not in, and comparison operators. Here's the list of the operators:

ATTRIBUTE	VALUE TYPE	DESCRIPTION
IN	any	Attribute value is member of a list.
NOT_IN	any	Attribute value is *not* member of a list.
== , >= , <= , > , <	int, float	Attribute value is equal, greater or equal, smaller or equal, greater or smaller.

Figure 4.2 – List of rich comparison operators

In our very first example, we matched good evening and good morning with two different patterns. Now, we can match good morning/evening with one pattern with the help of IN as follows:

```
doc = nlp("Good morning, I'm here. I'll say good evening!!")
pattern = [{"LOWER": "good"},
          {"LOWER": {"IN": ["morning", "evening"]}},
```

```
            {"IS_PUNCT": True}]
matcher.add("greetings",   [pattern])
matches = matcher(doc)
for mid, start, end in matches:
    print(start, end, doc[start:end])
...
0, 3, Good morning,
10, 13, good evening!
```

Comparison operators usually go together with the LENGTH attribute. Here's an example of finding long tokens:

```
doc = nlp("I suffered from Trichotillomania when I was in
college. The doctor prescribed me Psychosomatic medicine.")
pattern = [{"LENGTH": {">=" : 10}}]
matcher.add("longWords",   [pattern])
matches = matcher(doc)
for mid, start, end in matches:
    print(start, end, doc[start:end])
...
3, 4, Trichotillomania
14, 15, Psychosomatic
```

They were fun words to process! Now, we'll move onto another very practical feature of Matcher patterns, regex-like operators.

Regex-like operators

At the beginning of the chapter, we pointed out that spaCy's Matcher class offers a cleaner and more readable equivalent to regex operations, indeed much cleaner and much more readable. The most common regex operations are optional match (?), match at least once (+), and match 0 or more times (*). spaCy's Matcher also offers these operators by using the following syntax:

OP	DESCRIPTION
!	Negate the pattern, by requiring it to match exactly 0 times.
?	Make the pattern optional, by allowing it to match 0 or 1 times.
+	Require the pattern to match 1 or more times.
*	Allow the pattern to match zero or more times.

Figure 4.3 – OP key description

The very first example regex of this chapter was matching Barack Obama's first name, with the middle name being optional. The regex was as follows:

```
R"Barack\s(Hussein\s)?Obama
```

The ? operator after Hussein means the pattern in the brackets is optional, hence this regex matches both Barack Obama and Barack Hussein Obama. We use the ? operator in a Matcher pattern as follows:

```
doc1 = nlp("Barack Obama visited France.")
doc2 = nlp("Barack Hussein Obama visited France.")
pattern = [{"LOWER": "barack"},
           {"LOWER": "hussein", "OP": "?"},
           {"LOWER": "obama"}]
matcher.add("obamaNames", [pattern])
matcher(doc1)
[(1881848298847208418, 0, 2)]
matcher(doc2)
[(1881848298847208418, 0, 3)]
```

Here, by using the "OP": "?" in the second list item, we made this token optional. The matcher picked Barack Obama in the first doc object and Barack Hussein Obama in the second one as a result.

We previously pointed that the + and * operators have the same meaning as their regex counterparts. + means the token should occur at least once and * means the token can occur 0 or more times. Let's see some examples:

```
doc1 = nlp("Hello hello hello, how are you?")
doc2 = nlp("Hello, how are you?")
doc3 = nlp("How are you?")
```

```
pattern = [{"LOWER": {"IN": ["hello", "hi", "hallo"]},
            "OP":"*", {"IS_PUNCT": True}]
matcher.add("greetings",  [pattern])
for mid, start, end in matcher(doc1):
    print(start, end, doc1[start:end])
...
2, 4, hello,
1, 4, hello hello,
0, 4, Hello hello hello,
for mid, start, end in matcher(doc2):
    print(start, end, doc2[start:end])
...
0 2 Hello,

matcher(doc3)
...
[]
```

Here's what happened:

- In the pattern, the first token reads as *any one of hello, hi, hallo should occur 1 or more times* and the second token is punctuation.

- The third doc object does not match at all; there's no greeting word.

- The second doc object matches hello,.

When we come to the results of the first doc objects' matches, we see that there are not one, but three distinct matches. This is completely normal because there are indeed three sequences matching the pattern. If you have a closer look at the match results, all of them match the pattern we created, because hello, hello hello, and hello hello hello all match the (hello) + pattern.

Let's do the same pattern with * and see what happens this time:

```
doc1 = nlp("Hello hello hello, how are you?")
doc2 = nlp("Hello, how are you?")doc3 = nlp("How are you?")
pattern = [{"LOWER": {"IN": ["hello", "hi", "hallo"]},
            "OP": "+"}, {"IS_PUNCT": True}]
matcher.add("greetings",  [pattern])
for mid, start, end in matcher(doc1):
```

```
        print(start, end, doc1[start:end])
...
(0, 4, Hello hello hello,)
(1, 4, hello hello,)
(2, 4, hello,)
(3, 4, ,)
(7, 8, ?)
for mid, start, end in matcher(doc2):
        start, end, doc2[start:end]
...
(0, 2, hello,)
(1, 2, ,)
(5, 6, ?)
for mid, start, end in matcher(doc3):
        start, end, doc3[start:end]
...
(3, 4, ?)
```

In the first doc object's matches, there are two extra items: "" and ?. The "*" operator matches 0 or more, so our (hello)*punct_character pattern grabs "" and ?. The same applies to the second and third documents: punctuation marks alone without any greeting word are picked. This is not what you want in your NLP applications, probably.

The preceding example is a good example that we should be careful of while creating our patterns; sometimes, we get unwanted matches. For this reason, we usually consider using IS_SENT_START and take care with "*" operator.

The spaCy Matcher class also accepts a very special pattern, a **wildcard** token pattern. A wildcard token will match any token. We usually use it for the words we want to pick independent from their text or attributes or for the words we ignore. Let's see an example:

```
doc = nlp("My name is Alice and his name was Elliot.")
pattern = [{"LOWER": "name"},{"LEMMA": "be"},{}]
matcher.add("pickName", [pattern])
for mid, start, end in matcher(doc):
        print(start, end, doc[start:end])
...
1 4 name is Alice
6 9 name was Elliot
```

Here, we wanted to capture the first names in the sentence. We achieved it by parsing out token sequences in the form *name is/was/be firstname*. The first token pattern, LOWER: "name", matches the tokens whose lowered text is name. The second token pattern, LEMMA: "be", matches the is, was, and be tokens. The third token is the wildcard token, {}, which means *any* token. We pick up any token that comes after *name is/was/be* with this pattern.

We also use a wildcard token when we want to ignore a token. Let's make an example together:

```
doc1 = nlp("I forwarded his email to you.")
doc2 = nlp("I forwarded an email to you.")
doc3 = nlp("I forwarded the email to you.")
pattern = [{"LEMMA": "forward"}, {}, {"LOWER": "email"}]
matcher.add("forwardMail", [pattern])
for mid, start, end in matcher(doc1):
    print(start, end, doc1[start:end])
. . .
1 4 forwarded his email
for mid, start, end in matcher(doc2):
    print(start, end, doc2[start:end])
. . .
1 4 forwarded an email
for mid, start, end in matcher(doc3):
    print(start, end, doc3[start:end])
. . .
1 4 forwarded the email
```

It's just the opposite of the previous example. Here, we wanted to pick up *forward email* sequences, and we allowed that one token to come between forward and email. Here, the semantically important part is the forwarding an email action; whose email is it doesn't matter much.

We have mentioned regex quite a lot in this chapter so far, so now it's time to see how spaCy's Matcher class makes use of regex syntax.

Regex support

When we match individual tokens, usually we want to allow some variations, such as common typos, UK/US English character differences, and so on. Regex is very handy for this task and spaCy Matcher offers full support for token-level regex matching. Let's explore how we can use regex for our applications:

```
doc1 = nlp("I travelled by bus.")
doc2 = nlp("She traveled by bike.")
pattern = [{"POS": "PRON"},
           {"TEXT": {"REGEX": "[Tt]ravell?ed"}}]
for mid, start, end in matcher(doc1):
    print(start, end, doc1[start:end])
...
0 2 I traveled
for mid, start, end in matcher(doc2):
    print(start, end, doc2[start:end])
...
0 2 I travelled
```

Here, our second token pattern is [Tt]ravell?ed, which means the token can be capitalized or not. Also, there's an optional l after the first l. Allowing twin vowels and *ise/ize* alteration is a standard way of dealing with British and American English variations.

Another way of using regex is using it not only with text, but also with POS tags. What does the following code segment do?

```
doc = nlp("I went to Italy; he has been there too. His mother
also has told me she wants to visit Rome.")
pattern = [{"TAG": {"REGEX": "^V"}}]
matcher.add("verbs", [pattern])
for mid, start, end in matcher(doc):
    print(start, end, doc1[start:end])
...
1 2 went
6 7 has
7 8 been
14 15 has
15 16 told
```

```
18 19 wants
20 21 visit
```

We have extracted all the finite verbs (you can think of a finite verb as a non-modal verb). How did we do it? Our token pattern includes the regex ^V, which means all fine-grained POS tags that start with V: VB, VGD, VBG, VBN, VBP, and VBZ. Then we extracted tokens with verbal POS tags.

Looks tricky! Occasionally we use some tricks in NLU applications; while going through the examples of this book, you'll pick them up too. We encourage you to go over our examples and then try some example sentences of yours.

Matcher online demo

In the whole matching business, we occasionally see the match results visually. Regex offers *regex101* (`https://regex101.com/`), an online tool for checking if your regex pattern is working correctly (surprises with regex always happen). The following figure shows an example pattern and checking it against a text:

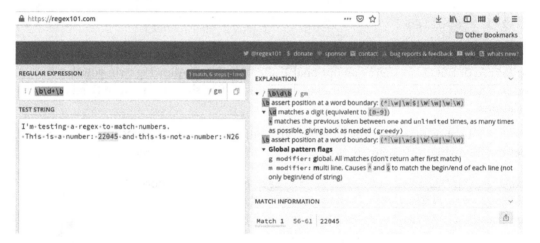

Figure 4.4 – An example regex match and pattern explanations

The explanations on the right side are quite detailed and illuminating. This is a tool used not only by NLP learners/beginners, but also professionals (regex can be quite difficult to read sometimes).

spaCy Matcher offers a similar tool on its online demo page (`https://explosion.ai/demos/matcher`). We can create patterns and test them against the text we want, interactively.

In the following screenshot, we can see a match example. On the right side we can select the attributes, values, and operators (such as +, *, !, and ?). After making this selection, the demo outputs the corresponding pattern string on the right side, below the checkboxes. On the left side, we first choose the spaCy language model we want (in this example, English core small), then see the results:

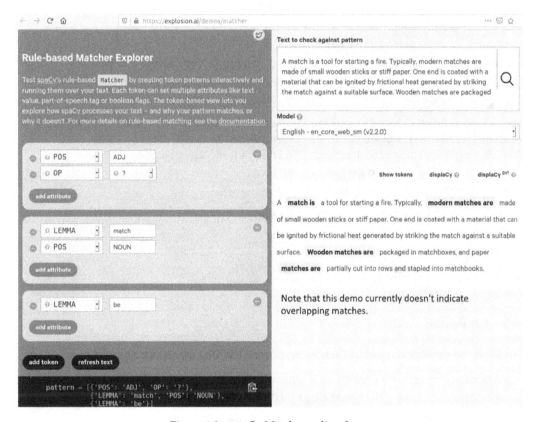

Figure 4.5 – spaCy Matcher online demo

Just like regex101, spaCy's Matcher demo helps you to see why your pattern matched or didn't match.

PhraseMatcher

While processing financial, medical, or legal text, often we have long lists and dictionaries and we want to scan the text against our lists. As we saw in the previous section, Matcher patterns are quite handcrafted; we coded each token individually. If you have a long list of phrases, Matcher is not very handy. It's not possible to code all the terms one by one.

spaCy offers a solution for comparing text against long dictionaries – the
`PhraseMatcher` class. The `PhraseMatcher` class helps us match long dictionaries.
Let's get started with an example:

```
import spacy
from spacy.matcher import PhraseMatcher
nlp = spacy.load("en_core_web_md")
matcher = PhraseMatcher(nlp.vocab)
terms = ["Angela Merkel", "Donald Trump", "Alexis Tsipras"]
patterns = [nlp.make_doc(term) for term in terms]
matcher.add("politiciansList", patterns)
doc = nlp("3 EU leaders met in Berlin. German chancellor Angela
Merkel first welcomed the US president Donald Trump. The
following day Alexis Tsipras joined them in Brandenburg.")
matches = matcher(doc)
for mid, start, end in matches:
    print(start, end, doc[start:end])
...
9 11 Angela Merkel
16 18 Donald Trump
22 24 Alexis Tsipras
```

Here's what we did:

- First, we imported `spacy`, then we imported the `PhraseMatcher` class.
- After the imports, we created a `Language` object, `nlp`, and initialized a
  `PhraseMatcher` object, `matcher`, with its vocabulary.
- The next two lines are where we created the pattern list.
- On line 6, we called `nlp.make_doc()` on the terms one by one to create the
  patterns.
- `make_doc()` creates a Doc from every term, and it's quite efficient in terms of
  processing because instead of the whole pipeline, it only calls the `Tokenizer`.
- The rest of the code is similar to what we did with Matcher: we iterated over the
  resulting spans.

This way, we match the pattern by their exact text values. What if we want to match them with other attributes? Here's an example of matching by the LOWER attribute:

```
matcher = PhraseMatcher(nlp.vocab, attr="LOWER")
terms = ["Asset", "Investment", "Derivatives",
         "Demand",  "Market"]
patterns = [nlp.make_doc(term) for term in terms]
matcher.add("financeTerms", patterns)
doc = nlp("During the last decade, derivatives market became an
asset class of their own and influenced the financial landscape
strongly.")
matches = matcher(doc)
for mid, start, end in matches:
    print(start, end, doc[start:end])
...
5 6 derivatives
6 7 market
```

On line 1, while creating a PhraseMatcher instance, we passed an additional argument, attr=LOWER. This way, the PhraseMatcher used the token.lower attribute during the match. Notice that the terms are uppercase and the matches are lowercase.

Another possible usage of PhraseMatcher is matching the SHAPE attribute. This matching strategy can be used on system logs, where IP numbers, dates, and other numerical values occur a lot. The good thing here is that you do not need to worry how the numbers are tokenized, you just leave it to PhraseMatcher. Let's see an example:

```
matcher = PhraseMatcher(nlp.vocab, attr="SHAPE")
ip_nums = ["127.0.0.0", "127.256.0.0"]
patterns = [nlp.make_doc(ip) for ip in ip_nums]
matcher.add("IPNums", patterns)
doc = nlp("This log contains the following IP addresses:
192.1.1.1 and 192.12.1.1 and 192.160.1.1 .")
for mid, start, end in matcher(doc):
    print(start, end, doc[start:end])
8 9 192.1.1.1
12 13 192.160.1.1
```

That's it! We matched the tokens and phrases successfully; what's left is named entities. Named entity extraction is an essential component of any NLP system and most of the pipelines you'll design will include an **named entity recognition** (**NER**) component. The next section is devoted to rule-based named entity extraction.

EntityRuler

While covering Matcher, we saw that we can extract named entities with Matcher by using the ENT_TYPE attribute. We recall from the previous chapter that ENT_TYPE is a linguistic attribute that refers to the entity type of the token, such as person, place, or organization. Let's see an example:

```
pattern = [{"ENT_TYPE": "PERSON"}]
matcher.add("personEnt", [pattern])
doc = nlp("Bill Gates visited Berlin.")
matches = matcher(doc)
for mid, start, end in matches:
    print(start, end, doc[start:end])
...
0 1 Bill
1 2 Gates
```

Again, we created a Matcher object called matcher and called it on the Doc object, doc. The result is two tokens, Bill and Gates; Matcher always matches at the token level. We got Bill and Gates, instead of the full entity, Bill Gates. If you want to get the full entity rather than the individual tokens, you can do this:

```
pattern = [{"ENT_TYPE": "PERSON", "OP": "+"}]
matcher.add("personEnt", [pattern])
doc = nlp("Bill Gates visited Berlin.")
matches = matcher(doc)
for mid, start, end in matches:
    print(start, end, doc[start:end])
...
0 1 Bill
1 2 Gates
0 2 Bill Gates
```

Usually, we match two or more entities together, or with other linguistic attributes to extract information. Here's an example of how we can understand the action in the sentence and which person in the sentence committed this action:

```
pattern = [{"ENT_TYPE": "PERSON", "OP": "+"}, {
                "POS" : "VERB"}]
matcher.add("personEntAction",  [pattern])
doc = nlp("Today German chancellor Angela Merkel met with the
US president.")
matches = matcher(doc)
for mid, start, end in matches:
    print(start, end, doc[start:end])
...
4 6 Merkel met
3 6 Angela Merkel met
```

We noticed that the Matcher returns two matches here; usually, we loop through the results and pick the longest match.

In the preceding examples, we matched the entities that the spaCy statistical model already extracted. What if we have domain-specific entities that we want to match? For instance, our dataset consists of wiki pages about ancient Greek philosophers. The philosopher names are naturally in Greek and don't follow English statistical patterns; it's expected that a tagger trained on English text would fail to extract the entity name occasionally. In these situations, we'd like spaCy to tell our entities and combine them with the statistical rules.

spaCy's `EntityRuler` is the component that allows us to add rules on top of the statistical model and creates an even more powerful **NER** model.

`EntityRuler` is not a matcher, it's a pipeline component that we can add to our pipeline via `nlp.add_pipe`. When it finds a match, the match is appended to `doc.ents` and `ent_type` will be the label we pass in the pattern. Let's see it in action:

```
doc = nlp("I have an acccount with chime since 2017")
doc.ents
(2017,)
patterns = [{"label": "ORG",
             "pattern": [{"LOWER": "chime"}]}]
ruler = nlp.add_pipe("entity_ruler")
ruler.add_patterns(patterns)
```

```
doc.ents
(chime, 2017)
doc[5].ent_type_
'ORG'
```

That's it, really easy, yet powerful! We added our own entity with just a couple of lines.

The `Matcher` class and `EntityRuler` are exciting and powerful features of the spaCy library, as we saw from the examples. Now, we move onto an exclusive section of quick and very handy recipes.

Combining spaCy models and matchers

In this section, we'll go through some recipes that will guide you through the entity extraction types you'll encounter in your NLP career. All the examples are ready-to-use and real-world recipes. Let's start with number-formatted entities.

Extracting IBAN and account numbers

IBAN and account numbers are two important entity types that occur in finance and banking frequently. We'll learn how to parse them out.

An IBAN is an international number format for bank account numbers. It has the format of a two-digit country code followed by numbers. Here are some IBANs from different countries:

Country	IBAN formatting example
Belgium	BE71 0961 2345 6769
Brazil	BR15 0000 0000 0000 1093 2840 814 P2
France	FR76 3000 6000 0112 3456 7890 189
Germany	DE91 1000 0000 0123 4567 89
Greece	GR96 0810 0010 0000 0123 4567 890
Mauritius	MU43 BOMM 0101 1234 5678 9101 000 MUR
Pakistan	PK70 BANK 0000 1234 5678 9000
Poland	PL10 1050 0099 7603 1234 5678 9123
Romania	RO09 BCYP 0000 0012 3456 7890
Saint Lucia	LC14 BOSL 1234 5678 9012 3456 7890 1234
Saudi Arabia	SA44 2000 0001 2345 6789 1234
Spain	ES79 2100 0813 6101 2345 6789
Switzerland	CH56 0483 5012 3456 7800 9
United Kingdom	GB98 MIDL 0700 9312 3456 78

Figure 4.6 – IBAN formats from different countries (source: Wikipedia)

How can we create a pattern for an IBAN? Obviously, in all cases, we start with two capital letters, followed by two digits. Then any number of digits can follow. We can express the country code and the next two digits as follows:

```
{"SHAPE": "XXdd"}
```

Here, XX corresponds to two capital letters and dd is two digits. Then XXdd pattern matches the first block of the IBAN perfectly. How about the rest of the digit blocks? For the rest of the blocks, we need to match a block of 1-4 digits. The regex \d{1,4} means a token consisting of 1 to 4 digits. This pattern will match a digit block:

```
{"TEXT": {"REGEX": "\d{1,4}"}}
```

We have a number of these blocks, so the pattern to match the digit blocks of an IBAN is as follows:

```
{"TEXT": {"REGEX": "\d{1,4}"}, "OP": "+"}
```

Then, we combine the first block with the rest of the blocks. Let's see the code and the matches:

```
doc = nlp("My IBAN number is BE71 0961 2345 6769, please send
the money there.")
doc1 = nlp("My IBAN number is FR76 3000 6000 0112 3456 7890
189, please send the money there.")
pattern = [{"SHAPE": "XXdd"},
          {"TEXT": {"REGEX": "\d{1,4}"}, "OP":"+"}]
matcher = Matcher(nlp.vocab)
matcher.add("ibanNum",  [pattern])
for mid, start, end in matcher(doc):
    print(start, end, doc[start:end])
...
4 6 BE71 0961
4 7 BE71 0961 2345
4 8 BE71 0961 2345 6769
for mid, start, end in matcher(doc1):
    print(start, end, doc1[start:end])
...
4 6 FR76 3000
4 7 FR76 3000 6000
4 8 FR76 3000 6000 0112
4 9 FR76 3000 6000 0112 3456
4 10 FR76 3000 6000 0112 3456 7890
4 11 FR76 3000 6000 0112 3456 7890 189
```

You can always follow a similar strategy when parsing numeric entities: first, divide the entity into some meaningful parts/blocks, then try to determine the shape or the length of the individual blocks.

We successfully parsed IBANs, now we can parse the account numbers. Parsing the account numbers is a bit trickier; account numbers are just plain numbers and don't have a special shape to help us differentiate them from usual numbers. What do we do, then? We can make a context lookup in this case; we can look around the number token and see if we can find *account number* or *account num* around the number token. This pattern should do the trick:

```
{"LOWER": "account"}, {"LOWER": {"IN": ["num", "number"]}},{},
{"IS_DIGIT": True}
```

We used a wildcard here: { } means any token. We allowed one token to go in between *number* and *account number*; this can be *is*, *was*, and so on. Let's see the code:

```
doc = nlp("My account number is 8921273.")
pattern = [{"LOWER": "account"},
            {"LOWER": {"IN": ["num", "number"]}},{},
            {"IS_DIGIT": True}]
matcher = Matcher(nlp.vocab)
matcher.add("accountNum", [pattern])
for mid, start, end in matcher(doc):
    print(start, end, doc[start:end])
...
1 5 account number is 8921273
```

If you want, you can include a possessive pronoun such as *my*, *your*, or *his* in the match, depending on the application's needs.

That's it for banking numbers. Now we'll extract another type of common numeric entity, phone numbers.

Extracting phone numbers

Phone numbers can have very different formats depending on the country, and matching phone numbers is often a tricky business. The best strategy here is to be specific about the country phone number format you want to parse. If there are several countries, you can add corresponding individual patterns to the matcher. If you have too many countries, then you can relax some conditions and go for a more general pattern (we'll see how to do that).

Let's start with the US phone number format. A US number is written as *(541) 754-3010* domestically or *+1 (541) 754-3010* internationally. We can form our pattern with an optional *+1*, then a three-digit area code, then two blocks of numbers separated with an optional *-*. Here is the pattern:

```
{"TEXT": "+1", "OP": "?"}, {"TEXT": "("}, {"SHAPE": "ddd"},
{"TEXT": ")"}, {"SHAPE": "ddd"}, {"TEXT": "-", "OP": "?"},
{"SHAPE": "dddd"}
```

Let's see an example:

```
doc1 = nlp("You can call my office on +1 (221) 102-2423 or
email me directly.")
```

```
doc2 = nlp("You can call me on (221) 102 2423 or text me.")
pattern = [{"TEXT": "+1", "OP": "?"}, {"TEXT": "("},
          {"SHAPE": "ddd"}, {"TEXT": ")"},
          {"SHAPE": "ddd"}, {"TEXT": "-", "OP": "?"},
          {"SHAPE": "dddd"}]
matcher = Matcher(nlp.vocab)
matcher.add("usPhonNum", [pattern])
for mid, start, end in matcher(doc1):
    print(start, end, doc1[start:end])
...
 6 13 +1 (221) 102-2423
for mid, start, end in matcher(doc2):
    print(start, end, doc2[start:end])
...
 5 11 (221) 102-2423
```

How about we make the pattern more general to apply to other countries as well? In this case, we can start with a 1 to 3-digit country code, followed by some digit blocks. It will match a broader set of numbers, so it's better to be careful not to match other numeric entities in your text.

We'll move onto textual entities from numeric entities. Now we'll process social media text and extract different types of entities that can occur in social media text.

Extracting mentions

Imagine analyzing a dataset of social media posts about companies and products and your task is to find out which companies are mentioned in which ways. The dataset will contain this sort of sentence:

```
CafeA is very generous with the portions.
CafeB is horrible, we waited for mins for a table.
RestaurantA is terribly expensive, stay away!
RestaurantB is pretty amazing, we recommend.
```

What we're looking for is most probably patterns of the *BusinessName is/was/be adverb* adjective* form. The following pattern would work:

```
[{"ENT_TYPE": "ORG"}, {"LEMMA": "be"}, {"POS": "ADV", "OP":
"*"}, {"POS": "ADJ"}]
```

Here, we look for an organization type entity, followed by a *is/was/be*, then optional adverbs, and finally an adjective.

What if you want to extract a specific business, let's say the company *ACME*? All you have to do is replace the first token with the specific company name:

```
[{"LOWER": "acme"}, {"LEMMA": "be"}, {"POS": "ADV", "OP": "*"},
{"POS": "ADJ"}]
```

That's it, easy peasy! After extracting the social media mentions, the next thing to do is to extract the hashtags and the emojis.

Hashtag and emoji extraction

Processing social media text is a hot topic and has some challenges. Social media text includes two sorts of unusual token types: hashtags and emojis. Both token types have a huge impact on the text meaning. The hashtag refers to the subject/object of the sentence, usually, and emojis can assign the sentiment of the sentence by themselves.

A hashtag consists of a # character at the beginning, then followed by a word of ASCII characters, with no inter-word spaces. Some examples are *#MySpace*, *#MondayMotivation* and so on. The spaCy tokenizer tokenizes these words into two tokens:

```
doc = nlp("#MySpace")
[token.text for token in doc]
['#', 'MySpace']
```

As a result, our pattern needs to match two tokens, the # character and the rest. The following pattern will match a hashtag easily:

```
{"TEXT": "#"}, {"IS_ASCII": True}
```

The following code extracts a hashtag:

```
doc = nlp("Start working out now #WeekendShred")
pattern = [{"TEXT": "#"}, {"IS_ASCII": True}]
matcher = Matcher(nlp.vocab)
```

```
matcher.add("hashTag",  [pattern])
matches = matcher(doc)
for mid, start, end in matches:
    print(start, end doc[start:end])
...
4 6 #WeekendShred
```

How about an emoji? An emoji is usually coded with lists according to their sentiment value, such as positive, negative, happy, sad, and so on. Here, we'll separate emojis into two classes, positive and negative. The following code spots the selected emoji in the text:

```
pos_emoji = ["😀", "😁", "😆", "😊", "😉", "😄"]
neg_emoji = ["😞", "😵", "😣", "😟", "😡", "😓"]
pos_patterns = [[{"ORTH": emoji}] for emoji in pos_emoji]
neg_patterns = [[{"ORTH": emoji}] for emoji in neg_emoji]
matcher = matcher(nlp.vocab)
matcher.add("posEmoji", pos_patterns)
matcher.add("negEmoji", neg_patterns)
doc = nlp(" I love Zara 😀 ")
for mid, start, end in matcher(doc):
    print(start, end, doc[start:end])
...
3 4 😀
```

Et voilà, the emoji 😀 is extracted happily! We'll make use of emojis in sentiment analysis chapter as well.

Now, let's extract some entities. We'll start with the common procedure of expanding named entities.

Expanding named entities

Often, we would like to expand a named entity's span to the left or to the right. Imagine you want to extract PERSON type named entities with titles so that you can deduce the gender or profession easily. spaCy's NER class already extracts person names, so how about the titles?

```
doc = nlp("Ms. Smith left her house 2 hours ago.")
doc.ents
(Smith, 2 hours ago)
```

As you see, the word Ms. is not included in the named entity because it's not a part of the person's name. A quick solution is to make a new entity type called TITLE:

```
patterns = [{"label": "TITLE", "pattern": [{"LOWER": {"IN":
["ms.", "mr.", "mrs.", "prof.", "dr."]}}]}]
ruler = nlp.add_pipe("entity_ruler")
ruler.add_patterns(patterns)
nlp.add_pipe(ruler)
doc = nlp("Ms. Smith left her house")
print([(ent.text, ent.label_) for ent in doc.ents])
[('Ms.', 'TITLE'), ('SMITH', 'PERSON')]
```

This is a quick and very handy recipe. You'll come across parsing titles a lot if you process wiki text or financial text.

In our next and final example, we'll combine POS attributes, dependency labels, and named entities.

Combining linguistic features and named entities

While charging meaning to a sentence, we evaluate word semantics by considering the contexts they occur in. Matching the words individually usually does not help us understand the full meaning. In most NLU tasks we have to combine linguistic features.

Imagine you're parsing professional biographies and make a work history of the subjects. You want to extract person names, the cities they have lived in, and the city they're currently working in.

Obviously we'll look for the word *live*; however, the POS tags hold the key here: whether it's the present tense or the past tense. In order to determine which city/place, we'll use syntactic information that is given by the dependency labels.

Let's examine the following example:

```
doc = nlp("Einstein lived in Zurich.")
[(ent.text, ent.label_) for ent in doc.ents]
[('Einstein', 'PERSON'), ('Zurich', 'GPE')]
```

Here is a visual representation of the preceding example:

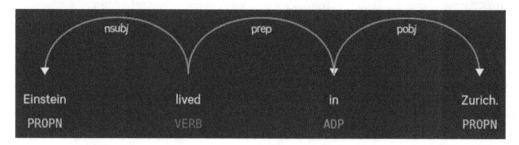

Figure 4.7 – Example parse, the entity "Einstein" being subject of the sentence

Here, lived is the main verb of the sentence, hence the root of the sentence. Einstein is the subject of the sentence, at the same time the person entity who lived. As we can see, the Einstein token's head is lived. There's also a place entity in the sentence, Zurich. If we follow the arcs from lived, we reach Zurich via a prepositional attachment. Finally, to determine the verb's tense, we can examine the POS tag. Let's see it in the following code:

```
person_ents = [ent for ent in doc.ents if ent.label_ ==
"PERSON"]
for person_ent in person_entities:
    #We use head of the entity's last token
    head = person_ent[-1].head
    If head.lemma_ == "live":
    #Check if the children of live contains prepositional
    attachment
    preps = [token for token in head.children if token.dep_ ==
"prep"]
    for prep in preps:
        places = [token for token in prep.children if token.
ent_type_ == "GPE"]
        # Verb is in past or present tense
        print({'person': person_ent, 'city': places,
            'past': head.tag_ == "VBD"})
```

Here, we combined POS tag information, dependency labels (hence syntactic information of the sentence), and named entities. It may not be easy for you to grasp it at first sight, but you'll get there by practicing.

Summary

This chapter introduced you to a very handy and powerful feature of spaCy, spaCy's matcher classes. You learned how to do rule-based matching with linguistic and token-level features. You learned about the `Matcher` class, spaCy's rule-based matcher. We explored the `Matcher` class by using it with different token features, such as shape, lemma, text, and entity type.

Then, you learned about `EntityRuler`, another lifesaving class that you can achieve a lot with. You learned how to extract named entities with the `EntityRuler` class.

Finally, we put together what you've learned in this chapter and your previous knowledge and combined linguistic features with rule-based matching with several examples. You learned how to extract patterns, entities of specific formats, and entities specific to your domain.

With this chapter, you completed the linguistic features. In the next chapter, we'll dive into the world of statistical semantics via a very important concept – **word vectors**. You'll discover the power of statistics in representing words, phrases, and sentences. Let's discover the world of semantics together!

5

Working with Word Vectors and Semantic Similarity

Word vectors are handy tools and have been the hot topic of NLP for almost a decade. A word vector is basically a dense representation of a word. What's surprising about these vectors is that semantically similar words have similar word vectors. Word vectors are great for semantic similarity applications, such as calculating the similarity between words, phrases, sentences, and documents. At a word level, word vectors provide information about synonymity, semantic analogies, and more. We can build semantic similarity applications by using word vectors.

Word vectors are produced by algorithms that make use of the fact that similar words appear in similar contexts. To capture the meaning of a word, a word vector algorithm collects information about the surrounding words that the target word appears with. This paradigm of capturing semantics for words by their surrounding words is called **distributional semantics**.

In this chapter, we will introduce the **distributional semantics paradigm** and its associated **semantic similarity methods**. We will start by taking a conceptual look at **text vectorization** so that you know what NLP problems word vectors solve.

Next, we will become familiar with word vector computations such as **distance calculation**, **analogy calculations**, and **visualization**. Then, we will learn how to benefit from spaCy's pretrained word vectors, as well as import and use third-party vectors. Finally, we will go through advanced semantic similarity methods using spaCy.

In this chapter, we're going to cover the following main topics:

- Understanding word vectors
- Using spaCy's pretrained vectors
- Using third-party word vectors
- Advanced semantic similarity methods

Technical requirements

In this chapter, we have used some external Python libraries besides spaCy for code visualization purposes. If you want to generate word vector visualizations in this chapter, you will need the following:

- NumPy
- scikit-learn
- Matplotlib

You can find this chapter's code in this book's GitHub repository: `https://github.com/PacktPublishing/Mastering-spaCy/tree/main/Chapter05`.

Understanding word vectors

The invention of word vectors (or **word2vec**) has been one of the most thrilling advancements in the NLP world. Those of you who are practicing NLP have definitely heard of word vectors at some point. This chapter will help you understand the underlying idea that caused the invention of word2vec, what word vectors look like, and how to use them in NLP applications.

The statistical world works with numbers, and all statistical methods, including statistical NLP algorithms, work with vectors. As a result, while working with statistical methods, we need to represent every real-world quantity as a vector, including text. In this section, we will learn about the different ways we can represent text as vectors and discover how word vectors provide semantic representation for words.

We will start by discovering text vectorization by covering the simplest implementation possible: one-hot encoding.

One-hot encoding

One-hot encoding is a simple and straightforward way to assign vectors to words: assign an index value to each word in the vocabulary and then encode this value into a **sparse vector**. Let's look at an example. Here, we will consider the vocabulary of a pizza ordering application; we can assign an index to each word in the order they appear in the vocabulary:

```
 1    a
 2    e-mail
 3    I
 4    cheese
 5    order
 6    phone
 7    pizza
 8    salami
 9    topping
10    want
```

Now, the vector of a vocabulary word will be 0, except for the position of the word's corresponding index value:

```
a          1 0 0 0 0 0 0 0 0 0
e-mail     0 1 0 0 0 0 0 0 0 0
I          0 0 1 0 0 0 0 0 0 0
cheese     0 0 0 1 0 0 0 0 0 0
order      0 0 0 0 1 0 0 0 0 0
phone      0 0 0 0 0 1 0 0 0 0
pizza      0 0 0 0 0 0 1 0 0 0
salami     0 0 0 0 0 0 0 1 0 0
topping    0 0 0 0 0 0 0 0 1 0
want       0 0 0 0 0 0 0 0 0 1
```

Now, we can represent a sentence as a matrix, where each row corresponds to one word. For example, the sentence *I want a pizza* can be represented by the following matrix:

```
I           0 0 1 0 0 0 0 0 0 0
want    0 0 0 0 0 0 0 0 0 1
a           1 0 0 0 0 0 0 0 0 0
pizza   0 0 0 0 0 0 1 0 0 0
```

As we can see from the preceding vocabulary and indices, the length of the vectors is equal to the number of the words in the vocabulary. Each dimension is devoted to one word explicitly. When we apply one-hot encoding vectorization to our text, each word is replaced by its vector and the sentence is transformed into a `(N, V)` matrix, where `N` is the number of words in the sentence and `V` is the vocabulary's size.

This way of representing text is straightforward to compute, as well as easy to debug and understand. This looks good so far, but there are some potential problems here, such as the following:

- The vectors are sparse. Each vector contains many 0s but only one `1`. Obviously, this is a waste of space if we know that words with similar meanings can be grouped together and share some dimensions. Also, numerical algorithms don't really like high-dimensional and sparse vectors in general.

- Secondly, what if the vocabulary size is over 1 million words? Obviously, we would need to use 1 million dimensional vectors, which is not really feasible in terms of memory and computation.

- Another problem is that the vectors are not *meaningful* at all. Similar words are not assigned similar vectors somehow. In the preceding vocabulary, the words `cheese`, `topping`, `salami`, and `pizza` actually carry related meanings, but their vectors are not related in any way. These vectors are indeed assigned randomly, depending on the corresponding word's index in the vocabulary. The one-hot encoded vectors don't capture any semantic relationships at all.

Word vectors were invented to answer the preceding list of concerns.

Word vectors

Word vectors are the solution to the preceding problems. A word vector is a **fixed-size**, **dense**, **real-valued** vector. From a broader perspective, a word vector is a learned representation of the text where semantically similar words have similar vectors. The following is what a word vector looks like. This has been extracted from **Glove English vectors** (we'll look at Glove in detail in the *How word vectors are produced* section):

```
the 0.418 0.24968 -0.41242 0.1217 0.34527 -0.044457 -0.49688
-0.17862 -0.00066023 -0.6566 0.27843 -0.14767 -0.55677 0.14658
-0.0095095 0.011658 0.10204 -0.12792 -0.8443 -0.12181 -0.016801
-0.33279 -0.1552 -0.23131 -0.19181 -1.8823 -0.76746 0.099051
-0.42125 -0.19526 4.0071 -0.18594 -0.52287 -0.31681 0.00059213
0.0074449 0.17778 -0.15897 0.012041 -0.054223 -0.29871 -0.15749
-0.34758 -0.045637 -0.44251 0.18785 0.0027849 -0.18411 -0.11514
-0.78581
```

This is a 50-dimensional vector for the word `the`. As you can see, the dimensions are floating points. But what do the dimensions represent? These individual dimensions typically don't have inherent meanings. Instead, they represent locations in the vector space, and the distance between these vectors indicates the similarity of the corresponding words' meanings. Hence, a word's meaning is distributed across the dimensions. This way of representing a word's meaning is called **distributional semantics**.

We've already mentioned that semantically similar words have similar representations. Let's look at the vectors of the different words and how they offer semantic representations. We can use the word vector visualizer for TensorFlow at https:// projector.tensorflow.org/ for this. On this website, Google offers word vectors for 10,000 words. Each vector is 200-dimensional and projected onto three dimensions for visualization. Let's look at the representation of the word cheese from our humble pizza ordering vocabulary:

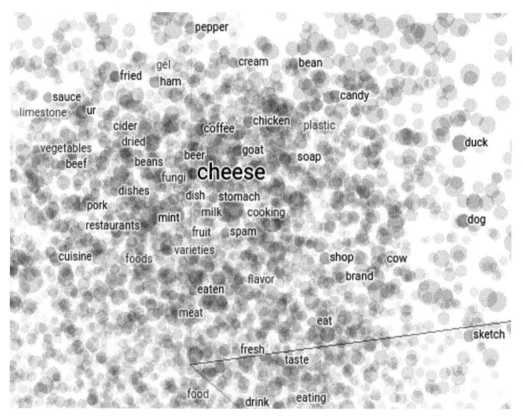

Figure 5.1 – The vector representation of the word "cheese" and semantically similar words

As we can see, the word cheese is semantically grouped with the other words about food. These are the words that are used together with the word cheese quite often: sauce, cola, food, and so on. In the following screenshot, we can see the closest words sorted by their cosine distance (think of cosine distance as a way of calculating the distance between vectors):

Figure 5.2 – Closest points to "cheese" in the three-dimensional space

How about some proper nouns? Word vectors are trained on a huge corpus, such as Wikipedia, which is why the representations of some proper nouns are also learned. For example, the proper noun **elizabeth** is represented by the following vector:

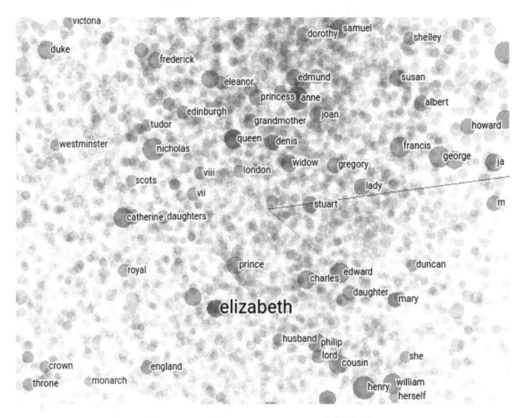

Figure 5.3 – Vector representation of elizabeth

> **Note**
>
> Notice that all the words in the preceding screenshot are in lowercase. Most of the word vector algorithms make all the vocabulary input words lowercase to avoid there being two representations of the same word.

Here, we can see that **elizabeth** indeed points to Queen Elizabeth of England. The surrounding words are **monarch, empress, princess, royal, lord, lady, crown, England, Tudor, Buckingham,** her mother's name, **anne,** her father's name, **henry,** and even her mother's rival queen's name, **catherine**! Both ordinary words such as **crown** and proper nouns such as **henry** are grouped together with **elizabeth**. We can also see that the syntactic category of all the neighbor words is noun; verbs don't go together with nouns.

Word vectors can capture synonyms, antonyms, and semantic categories such as animals, places, plants, names, and abstract concepts. Next, we'll dive deep into semantics and explore a surprising feature provided by word vectors – **word analogies**.

Analogies and vector operations

We've already seen that learned representations can capture semantics. What's more, word vectors support vector operations, such as vector addition and subtraction, in a meaningful way. Indeed, adding and subtracting word vectors is one way to support analogies.

A word analogy is a semantic relationship between a pair of words. There are many types of relationship, such as synonymity, anonymity, and wholepart relation. Some example pairs are (King – man, Queen – woman), (airplane – air, ship - sea), (fish – sea, bird - air), (branch – tree, arm – human), (forward – backward, absent – present), and so on.

For example, we can represent gender mapping between the Queen and King as Queen – Woman + Man = King. Here, if we subtract *woman* from *Queen* and add *man* instead, we get *King*. Then, this analogy reads as, *queen is to king as woman is to man*. Embeddings can generate remarkable analogies such as gender, tense, and capital city. The following diagram shows these analogies:

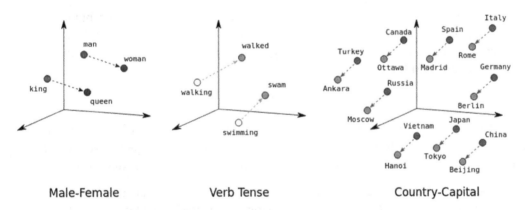

| Male-Female | Verb Tense | Country-Capital |

Figure 5.4 – Analogies created by the word vectors (Source: https://developers.google.com/machine-learning/crash-course/embeddings/translating-to-a-lower-dimensional-space)

Obviously, word vectors provide great semantic capabilities for NLP developers, but how are they produced? We'll learn more about word vector generation algorithms in the next section.

How word vectors are produced

There is more than one way to produce word vectors. Let's look at the most popular pretrained vectors and how they are trained:

- **word2vec** is the name of the statistical algorithm that was created by Google to produce word vectors. Word vectors are trained with a neural network architecture, which processes windows of words and predicts the vector for each word, depending on the surrounding words. These pretrained word vectors can be downloaded from `https://developer.syn.co.in/tutorial/bot/oscova/pretrained-vectors.html#word2vec-and-glove-models`. We won't go into the details here, but you can read the excellent blog at `https://jalammar.github.io/illustrated-word2vec/` for more details about the algorithm and data preparation steps.

- **Glove** vectors are trained in another way and were invented by the Stanford NLP group. This method depends on singular value decomposition, which is used on the word co-occurrences matrix. A comprehensive guide to the Glove algorithm is available at `https://www.youtube.com/watch?v=Fn_U2OG1uqI`. The pretrained vectors are available at `https://nlp.stanford.edu/projects/glove/`.

- **fastText** was created by Facebook Research and is similar to word2vec, but offers more. word2vec predicts words based on their surrounding context, while fastText predicts subwords; that is, character n-grams. For example, the word *chair* generates the following subwords:

```
ch, ha, ai, ir, cha, hai, air
```

fastText produces a vector for each subword, including misspelled words, numbers, partial words, and single characters. fastText is robust when it comes to misspelled words and rare words. It can compute a vector for the tokens that are not proper lexicon words.

Facebook Research published pretrained fastText vectors for 157 languages. You can find these models at `https://fasttext.cc/docs/en/crawl-vectors.html`

All the preceding algorithms follow the same idea: similar words occur in a similar context. The context – the surrounding words around a word – is key to generating the word vector for a specific word in any case. All the pretrained word vectors that are generated with the preceding three algorithms are trained on a huge corpus such as Wikipedia, the news, or Twitter.

> **Pro tip**
>
> When we say similar words, the first concept that comes to mind is
> **synonymity**. Synonym words occur in a similar context; for example, *freedom*
> and *liberty* both mean the same thing:
>
> We want free healthcare, education, and liberty.
>
> We want free healthcare, education, and freedom.
>
> How about antonyms? Antonyms can be used in the same context. Take *love*
> and *hate*, for example:
>
> I hate cats.
>
> I love cats.
>
> As you can see, antonyms also appear in similar contexts; hence, usually, their
> vectors are also similar. If your downstream NLP task is sensitive in this aspect,
> be careful while using word vectors. In this case, always either train your
> own vectors or refine your word vectors by training them in the downstream
> task as well. You can train your own word vectors with the Gensim package
> (`https://radimrehurek.com/gensim/`). The Keras library allows
> word vectors to be trained on downstream tasks. We'll revisit this issue in
> *Chapter 8, Text Classification with spaCy*.

Now that we know more about word vectors, let's look at how to use spaCy's pretrained
word vectors.

Using spaCy's pretrained vectors

We installed a medium-sized English spaCy language model in *Chapter 1, Getting
Started with spaCy*, so that we can directly use word vectors. Word vectors are part of
many spaCy language models. For instance, the `en_core_web_md` model ships with
300-dimensional vectors for 20,000 words, while the `en_core_web_lg` model ships
with 300-dimensional vectors with a 685,000 word vocabulary.

Typically, small models (those whose names end with `sm`) do not include any word vectors
but include context-sensitive tensors. You can still make the following semantic similarity
calculations, but the results won't be as accurate as word vector computations.

You can reach a word's vector via the `token.vector` method. Let's look at this method in an example. The following code queries the word vector for banana:

```
import spacy
nlp = spacy.load("en_core_web_md")
doc = nlp("I ate a banana.")
doc[3].vector
```

The following screenshot was taken within the Python shell:

```
>>> doc[3].vector
array([ 2.0228e-01, -7.6618e-02,  3.7032e-01,  3.2845e-02, -4.1957e-01,
        7.2069e-02, -3.7476e-01,  5.7460e-02, -1.2401e-02,  5.2949e-01,
       -5.2380e-01, -1.9771e-01, -3.4147e-01,  5.3317e-01, -2.5331e-02,
        1.7380e-01,  1.6772e-01,  8.3984e-01,  5.5107e-02,  1.0547e-01,
        3.7872e-01,  2.4275e-01,  1.4745e-02,  5.5951e-01,  1.2521e-01,
       -6.7596e-01,  3.5842e-01, -4.0028e-02,  9.5949e-02, -5.0690e-01,
       -8.5318e-02,  1.7980e-01,  3.3867e-01,  1.3230e-01,  3.1021e-01,
        2.1878e-01,  1.6853e-01,  1.9874e-01, -5.7385e-01, -1.0649e-01,
        2.6669e-01,  1.2838e-01, -1.2803e-01, -1.3284e-01,  1.2657e-01,
        8.6723e-01,  9.6721e-02,  4.8306e-01,  2.1271e-01, -5.4990e-02,
       -8.2425e-02,  2.2408e-01,  2.3975e-01, -6.2260e-02,  6.2194e-01,
       -5.9900e-01,  4.3201e-01,  2.8143e-01,  3.3842e-02, -4.8815e-01,
       -2.1359e-01,  2.7401e-01,  2.4095e-01,  4.5950e-01, -1.8605e-01,
       -1.0497e+00, -9.7305e-02, -1.8908e-01, -7.0929e-01,  4.0195e-01,
       -1.8768e-01,  5.1687e-01,  1.2520e-01,  8.4150e-01,  1.2097e-01,
        8.8239e-02, -2.9196e-02,  1.2151e-03,  5.6825e-02, -2.7421e-01,
        2.5564e-01,  6.9793e-02, -2.2258e-01, -3.6006e-01, -2.2402e-01,
       -5.3699e-02,  1.2022e+00,  5.4535e-01, -5.7998e-01,  1.0905e-01,
        4.2167e-01,  2.0662e-01,  1.2936e-01, -4.1457e-02, -6.6777e-01,
        4.0467e-01, -1.5218e-02, -2.7640e-01, -1.5611e-01, -7.9198e-02,
        4.0037e-02, -1.2944e-01, -2.4090e-04, -2.6785e-01, -3.8115e-01,
       -9.7245e-01,  3.1726e-01, -4.3951e-01,  4.1934e-01,  1.8353e-01,
       -1.5260e-01, -1.0808e-01, -1.0358e+00,  7.6217e-02,  1.6519e-01,
        2.6526e-04,  1.6616e-01, -1.5281e-01,  1.8123e-01,  7.0274e-01,
        5.7956e-03,  5.1664e-02, -5.9745e-02, -2.7551e-01, -3.9049e-01,
        6.1132e-02,  5.5430e-01, -8.7997e-02, -4.1681e-01,  3.2826e-01,
       -5.2549e-01, -4.4288e-01,  8.2183e-03,  2.4486e-01, -2.2982e-01,
       -3.4981e-01,  2.6894e-01,  3.9166e-01, -4.1904e-01,  1.6191e-01,
       -2.6263e+00,  6.4134e-01,  3.9743e-01, -1.2868e-01, -3.1946e-01,
       -2.5633e-01, -1.2220e-01,  3.2275e-01, -7.9933e-02, -1.5348e-01,
        3.1505e-01,  3.0591e-01,  2.6012e-01,  1.8553e-01, -2.4043e-01,
```

Figure 5.5 – Word vector for the word "banana"

`token.vector` returns a NumPy `ndarray`. You can call numpy methods on the result:

```
type(doc[3].vector)
<class 'numpy.ndarray'>
doc[3].vector.shape
(300,)
```

In this code segment, first, we queried the Python type of the word vector. Then, we invoked the `shape()` method of the NumPy array on the vector.

The `Doc` and `Span` objects also have vectors. The vector of a sentence or a span is the average of its words' vectors. Run the following code and view the results:

```
doc = nlp("I like a banana,")
doc.vector
doc[1:3].vector
```

Only the words in the model's vocabulary have vectors; words that are not in the vocabulary are called **OOV (out-of-vocabulary)** words. `token.is_oov` and `token.has_vector` are two methods we can use to query whether a token is in the model's vocabulary and has a word vector:

```
doc = nlp("You went there afskfsd.")
for token in doc:
            token.is_oov, token.has_vector
(False, True)
(False, True)
(False, True)
(True, False)
(False, True)
```

This is basically how we use spaCy's pretrained word vectors. Next, we'll discover how to invoke spaCy's semantic similarity method on `Doc`, `Span`, and `Token` objects.

The similarity method

In spaCy, every container type object has a similarity method that allows us to calculate the semantic similarity of other container objects by comparing their word vectors.

We can calculate the semantic similarity between two container objects, even though they are different types of containers. For instance, we can compare a `Token` object to a `Doc` object and a `Doc` object to a `Span` object. The following example computes how similar two `Span` objects are:

```
doc1 = nlp("I visited England.")
doc2 = nlp("I went to London.")
doc1[1:3].similarity(doc2[1:4])
0.6539691
```

We can compare the two `Token` objects, `London` and `England`, as well:

```
doc1[2].similarity(doc2[3])
0.73891276
```

The sentence's similarity is computed by calling `similarity()` on the `Doc` objects:

```
doc1.similarity(doc2)
0.7995623615797786
```

The preceding code segment calculates the semantic similarity between the two sentences I visited England. and I went to London.. The similarity score is high enough that it considers both sentences are similar (the degree of similarity ranges from 0 to 1, with 0 being unrelated and 1 being identical).

Not surprisingly, the `similarity()` method returns 1 when you compare an object to itself:

```
doc1.similarity(doc1)
1.0
```

Judging the distance with numbers is difficult sometimes, but looking at the vectors on paper can also help us understand how our vocabulary words are grouped. The following code snippet visualizes a simple vocabulary of two semantic classes. The first class of words is for animals, while the second class is for food. We expect these two classes of words to become two groups on the graphics:

```
import matplotlib.pyplot as plt
from sklearn.decomposition import PCA
import numpy as np
import spacy
nlp = spacy.load("en_core_web_md")
```

```
vocab = nlp("cat dog tiger elephant bird monkey lion cheetah
burger pizza food cheese wine salad noodles macaroni fruit
vegetable")
words = [word.text for word in vocab]
>>> vecs = np.vstack([word.vector for word in vocab if word.
has_vector])
pca = PCA(n_components=2)
vecs_transformed = pca.fit_transform(vecs)
plt.figure(figsize=(20,15))
plt.scatter(vecs_transformed[:,0], vecs_transformed[:,1])
for word, coord in zip(words, vecs_transformed):
        x,y = coord
        plt.text(x,y,word, size=15)
plt.show()
```

This code snippet achieves a lot. Let's take a look:

1. First, we imported the matplotlib library for creating our graphic.

2. The next two imports are for calculating the vectors.

3. We imported `spacy` and created an `nlp` object as usual.

4. Then, we created a `Doc` object from our vocabulary.

5. Next, we stacked our word vectors vertically by calling `np.vstack`.

6. Since the vectors are 300-dimensional, we needed to project them into a two-dimensional space for visualization purposes. We made this projection by extracting the two principal components via **principal component analysis (PCA)**.

7. The rest of the code deals with matplotlib function calls to create a scatter plot.

The resulting visual looks as follows:

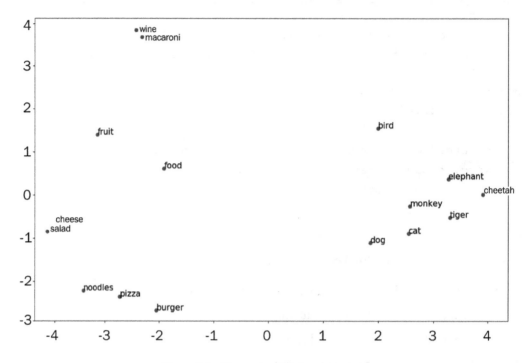

Figure 5.6 – Two semantic classes grouped

Voilà! Our spaCy word vectors really worked! Here, we can see the two semantic classes that were grouped on the visualization. Notice that the distance between the animals is less and more uniformly distributed, while the food class formed groups inside the group.

Previously, we mentioned that we can create our own word vectors or refine them on our own corpus. Once we've done that, can we use them within spaCy? The answer is yes! In the next section, we'll learn how to load custom word vectors into spaCy.

Using third-party word vectors

We can also use third-party word vectors within spaCy. In this section, we'll learn how to import a third-party word vector package into spaCy. We'll use fastText's subword-based pretrained vectors from the Facebook AI. You can view the list of all the available English pretrained vectors at https://fasttext.cc/docs/en/english-vectors.html.

The name of the package identifies the vector's dimension, the vocabulary size, and the corpus genre that the vectors will be trained on. For instance, `wiki-news-300d-1M-subword.vec.zip` indicates that it contains 1 million 300-dimensional word vectors that have been trained on a Wikipedia corpus.

Let's start downloading the vectors:

1. In your terminal, type the following command. Alternatively, you can copy and paste the URL into your browser and the download should start:

   ```
   $ wget https://dl.fbaipublicfiles.com/fasttext/vectors-
   english/wiki-news-300d-1M-subword.vec.zip
   ```

   The preceding line will download the 300-dimensional word vectors onto your machine.

2. Next, we will unzip the following `.zip` file. You can either unzip it by right-clicking or by using the following code:

   ```
   $ unzip wiki-news-300d-1M-subword.vec.zip
   ```

   After unzipping this file, you should see the `wiki-news-300d-1M-subword.vec` file.

3. Now, we're ready to use spaCy's `init-model` command:

   ```
   $ $ python -m spacy init-model en en_subwords_wiki_lg
   --vectors-loc wiki-news-300d-1M-subword.vec
   ```

   This command performs the following actions:

   a) Converts the `wiki-news-300d-1M-subword.vec` vectors into spaCy's vector format.

   b) Creates a language model directory named en_subwords_wiki_lg that contains the newly created vectors.

4. If everything goes well, you should see the following message:

   ```
   Reading vectors from wiki-news-300d-1M-subword.vec
   Open loc
   999994it [02:05, 7968.84it/s]
   Creating model...
   0it [00:00, ?it/s]        Successfully compiled vocab
         999731 entries, 999994 vectors
   ```

5. With that, we've created the language model. Now, we can load it:

```python
import spacy
nlp = spacy.load("en_subwords_wiki_lg")
```

6. Now, we can create a `doc` object with this `nlp` object, just like we did with spaCy's default language models:

```python
doc = nlp("I went there.")
```

The model we just created is an empty model that was initiated with the word vectors, so it does not contain any other pipeline components. For instance, making a call to `doc.ents` will fail with an error. So, be careful while working with third-party vectors and favor built-in spaCy vectors whenever possible.

Advanced semantic similarity methods

In this section, we'll discover advanced semantic similarity methods for word, phrase, and sentence similarity. We've already learned how to calculate semantic similarity with spaCy's **similarity** method and obtained some scores. But what do these scores mean? How are they calculated? Before we look at more advanced methods, first, we'll learn how semantic similarity is calculated.

Understanding semantic similarity

When we collect text data (any sort of data), we want to see how some examples are similar, different, or related. We want to measure how similar two pieces of text are by calculating their similarity scores. Here, the term *semantic similarity* comes into the picture; **semantic similarity** is a **metric** that's defined over texts, where the distance between two texts is based on their semantics.

A metric in mathematics is basically a distance function. Every metric induces a topology on the vector space. Word vectors are vectors, so we want to calculate the distance between them and use this as a similarity score.

Now, we'll learn about two commonly used distance functions: **Euclidian distance** and **cosine distance**. Let's start with Euclidian distance.

Euclidian distance

The Euclidian distance between two points in a k-dimensional space is the length of the path between them. The distance between two points is calculated by the Pythagorean theorem. We calculate this distance by summing the difference of each coordinate's square and then taking the square root of this sum. The following diagram shows the Euclidian distance between two vectors, dog and cat:

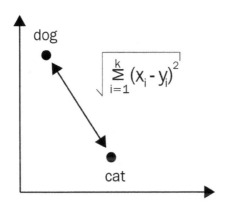

Figure 5.7 – Euclidian distance between two vectors, dog and cat

What does Euclidian distance mean for word vectors? First, Euclidian distance has no idea of **vector orientation**; what matters is the **vector magnitude**. If we take a pen and draw a vector from the origin to the **dog** point (let's call it **dog vector**) and do the same for the **cat** point (let's call it **cat vector**) and subtract one vector from and other, then the distance is basically the magnitude of this difference vector.

What happens if we add two more semantically similar words (*canine, terrier*) to **dog** and make it a text of three words? Obviously, the dog vector will now grow in magnitude, possibly in the same direction. This time, the distance will be much bigger due to geometry (as shown in the following diagram), although the semantics of the first piece of text (now **dog canine terrier**) remain the same.

This is the main drawback of using Euclidian distance for semantic similarity – the orientation of the two vectors in the space is not taken into account. The following diagram illustrates the distance between **dog** and **cat** and the distance between **dog canine terrier** and **cat**:

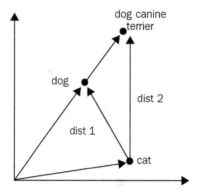

Figure 5.8 – Distance between "dog" and "cat," as well as the distance between "dog canine terrier" and "cat"

How can we fix this problem? There's another way of calculating similarity that addresses this problem, called **cosine similarity**. Let's take a look.

Cosine distance and cosine similarity

Contrary to Euclidian distance, cosine distance is more concerned with the orientation of the two vectors in the space. The cosine similarity of two vectors is basically the cosine of the angle that's created by these two vectors. The following diagram shows the angle between the **dog** and **cat** vectors:

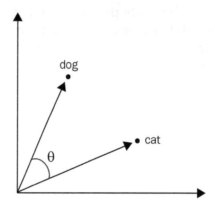

Figure 5.9 – The angle between the dog and cat vectors. Here, the semantic similarity is calculated by $\cos(\theta)$

The maximum similarity score that's allowed by cosine similarity is 1. This is obtained when the angle between two vectors is 0 degrees (hence, the vectors coincide). The similarity between two vectors is 0 when the angle between them is 90 degrees.

Cosine similarity provides us with scalability when the vectors grow in magnitude. We will refer to *Figure 5.8* again here. If we grow one of the input vectors, the angle between them remains the same, so the cosine similarity score is the same.

Note that here, we are calculating the semantic similarity score, not the distance. The highest possible value is 1 when the vectors coincide, while the lowest score is 0 when two vectors are perpendicular. The cosine distance is $1 - \cos(\theta)$, which is a distance function.

spaCy uses cosine similarity to calculate semantic similarity. Hence, calling the `similarity` method helps us make cosine similarity calculations.

So far, we've learned how to calculate similarity scores, but we still haven't discovered words we should look for meaning in. Obviously, not all the words in a sentence have the same impact on the semantics of the sentence. The similarity method will calculate the semantic similarity score for us, but for the results of that calculation to be useful, we need to choose the right keywords to compare. To understand why, consider the following text snippet:

```
Blue whales are the biggest mammals in the world. They're
observed in California coast during spring.
```

If we're interested in finding the biggest mammals on the planet, the phrases `biggest mammals` and `in the world` will be the key words. Comparing these phrases with the search phrases *largest mammals* and *on the planet* should give us a high similarity score. But if we're interested in finding out about some places in the world, `California` will be the keyword. `California` is semantically similar to the word *geography* and, even better, the entity type is a geographical noun.

We have already learned *how* to calculate the similarity score. In the next section, we'll learn about *where* to look for the *meaning*. We'll extract the key phrases and named entities from the sentences and then use them in similarity score calculations. We'll start by covering a case study on text categorization before improving the task results via key phrase extraction.

Categorizing text with semantic similarity

Determining two sentence's semantic similarity can help you categorize texts into predefined categories or spot only the relevant texts. In this case study, we'll filter all user comments in an e-commerce website related to the word *perfume*. Suppose you need to evaluate the following user comments:

```
I purchased a science fiction book last week.
I loved everything related to this fragrance: light, floral and
feminine …
I purchased a bottle of wine.
```

Here, we can see that only the second sentence is related. This is because it contains the word `fragrance`, as well as the adjectives describing scents. To understand which sentences are related, we can try several comparison strategies.

First, we can compare `perfume` to each sentence. Recall that spaCy generates a word vector for a sentence by averaging the word vector of its tokens. The following code snippet compares the preceding sentences to the `perfume` search key:

```
sentences = nlp("I purchased a science fiction book last week.
I loved everything related to this fragrance: light, floral and
feminine... I purchased a bottle of wine.  ")
key = nlp("perfume")
for sent in sentences.sents:
            print(sent.similarity(key))
...
0.2481654331382154
0.5075297559861377
0.42154297167069865
```

Here, we performed the following steps:

1. First, we created a `Doc` object with the three preceding sentences.

2. Then, for each sentence, we calculated the similarity score with `perfume`.

3. Then, we printed the score by invoking the `similarity()` method on the sentence.

The degree of similarity between `perfume` and the first sentence is small, indicating that this sentence is not very relevant to our search key. The second sentence looks relevant, which means that we correctly spotted the semantic similarity.

How about the third sentence? The script identified that the third sentence is relevant somehow, most probably because it includes the word `bottle` and perfumes are sold in bottles. The word `bottle` appears in similar contexts with the word `perfume`. For this reason, the similarity score of this sentence and the search key is not low enough; also, the scores of the second sentence and the third sentence are not far away enough to make the second sentence significant.

There's another potential problem with comparing the key to the whole sentence. In practice, we occasionally deal with quite long texts, such as web documents. Averaging over a very long text lowers the importance of key words.

To improve performance, we can extract the *important* words. Let's look at how we can spot the key phrases in a sentence.

Extracting key phrases

A better way to do semantic categorization is to extract the important words/phrases and compare only them to the search key. Instead of comparing the key to the different parts of speech, we can compare the key to just the noun phrases. Noun phrases are the subjects, direct objects, and indirect objects of the sentences and carry a big percentage of the sentence's semantics on their shoulders.

For example, in the sentence *Blue whales live in California.*, you'd probably like to focus on *blue whales, whales, California,* or *whales in California*.

Similarly, in the preceding sentence about perfume, we focused on picking out the noun, *fragrance*. In different semantic tasks, you might need other context words such as verbs to decide what the sentence is about, but for semantic similarity, noun phrases carry most weight.

What is a noun phrase, then? A **noun phrase** (**NP**) is a group of words that consist of a noun and its modifiers. Modifiers are usually pronouns, adjectives, and determiners. The following phrases are noun phrases:

```
A dog
My dog
My beautiful dog
A beautiful dog
A beautiful and happy dog
My happy and cute dog
```

spaCy extracts noun phases by parsing the output of the dependency parser. We can see the noun phrases of a sentence by using the doc.noun_chunks method:

```
doc = nlp("My beautiful and cute dog jumped over the fence")
doc.noun_chunks
<generator object at 0x7fa3c529be58>
list(doc.noun_chunks)
[My beautiful and cute dog, the fence]
```

Let's modify the preceding code snippet a bit. Instead of comparing the search key *perfume* to the entire sentence, this time, we will only compare it with the sentence's noun chunks:

```
for sent in sentences.sents:
        nchunks = [nchunk.text for nchunk in sent.noun_
chunks]
        nchunk_doc = nlp(" ".join(nchunks))
        print(nchunk_doc.similarity(key))
0.21390893517254456
0.6047741393523175
0.44506391511570403
```

In the preceding code, we did the following:

1. First, we iterated over the sentences.

2. Then, for each sentence, we extracted the noun chunks and stored them in a Python list.

3. Next, we joined the noun chunks in the list into a Python string and converted it into a Doc object.

4. Finally, we compared this Doc object of noun chunks to the search key *perfume* to determine their semantic similarity score.

If we compare these scores to the previous scores, we will see that the first sentence is still irrelevant, so its score went down slightly. The second sentence's score increased significantly. Now, the second sentence's and the third sentence's scores look so far away from each other for us to confidently say that the second sentence is the most related sentence here.

Extracting and comparing named entities

In some cases, instead of extracting every noun, we will only focus on the proper nouns; hence, we want to extract the named entities. Let's say we want to compare the following paragraphs:

> "Google Search, often referred as Google, is the most popular search engine nowadays. It answers a huge volume of queries every day."

> "Microsoft Bing is another popular search engine. Microsoft is known by its star product Microsoft Windows, a popular operating system sold over the world."

> "The Dead Sea is the lowest lake in the world, located in the Jordan Valley of Israel. It is also the saltiest lake in the world."

Our code should be able to recognize that the first two paragraphs are about large technology companies and their products, while the third paragraph is about a geographic location.

Comparing all the noun phrases in these sentences may not be very helpful because many of them, such as volume, aren't relevant to the categorization. The topics of these paragraphs are determined by the phrases within them; that is, Google Search, Google, Microsoft Bing, Microsoft, Windows, Dead Sea, Jordan Valley, and Israel. spaCy can spot these entities:

```
doc1 = nlp("Google Search, often referred as Google, is the
most popular search engine nowadays. It answers a huge volume
of queries every day.")
doc2 = nlp("Microsoft Bing is another popular search engine.
Microsoft is known by its star product Microsoft Windows, a
popular operating system sold over the world.")
doc3 = nlp("The Dead Sea is the lowest lake in the world,
located in the Jordan Valley of Israel. It is also the saltiest
lake in the world.")
doc1.ents
(Google,)
doc2.ents
(Microsoft Bing, Microsoft, Microsoft, Windows)
doc3.ents
(The Dead Sea, the Jordan Valley, Israel)
```

Now that we have extracted the words we want to compare, let's calculate the similarity scores:

```
ents1 = [ent.text for ent in doc1.ents]
ents2 = [ent.text for ent in doc2.ents]
ents3 = [ent.text for ent in doc3.ents]
ents1 = nlp(" ".join(ents1))
ents2 = nlp(" ".join(ents2))
ents3 = nlp(" ".join(ents3))
ents1.similarity(ents2)
0.6078712596225045
ents1.similarity(ents3)
0.374100398233877
ents2.similarity(ents3)
0.36244710903224026
```

Looking at these figures, we can see that the highest level of similarity exists between the first and the second paragraph, which are both about large tech companies. The third paragraph is not really similar to the other paragraphs. How did we get this calculation by just using word vectors? Probably because the words *Google* and *Microsoft* often appear together in news and other social media text corpuses, hence creating similar word vectors.

Congratulations! You've reached the end of the *Advanced semantic similarity methods* section! You explored different ways of combining word vectors with linguistic features such as key phrases and named entities. By finishing this section, we are now ready to conclude this chapter.

Summary

In this chapter, you worked with word vectors, which are floating-point vectors that represent word semantics. First, you learned about the different ways to perform text vectorization, as well as how to use word vectors and distributed semantics. Then, you explored the vector operations that word vectors allow and what semantics these operations bring.

You also learned how to use spaCy's built-in word vectors and how to import third-party vectors into spaCy. Finally, you learned about vector-based semantic similarity and how to blend linguistic concepts with word vectors to get the best out of these semantics.

The next chapter is full of surprises – we'll look at a real-word case-based study that allows you to blend what you've learned about in the past five chapters. Let's see what spaCy can do when it comes to real-world problems!

6
Putting Everything Together: Semantic Parsing with spaCy

This is a purely hands-on section. In this chapter, we will apply what we have learned hitherto to **Airline Travel Information System (ATIS)**, a well-known airplane ticket reservation system dataset. First of all, we will get to know our dataset and make the basic statistics. As the first **natural language understanding (NLU)** task, we will extract the named entities with two different methods, with spaCy Matcher, and by walking on the dependency tree.

The next task is to determine the intent of the user utterance. We will explore intent recognition in different ways, too: by extracting the verbs and their direct objects, by using wordlists, and by walking on the dependency tree to recognize multiple intents. Then you will match your keywords to synonyms from a synonyms list to detect semantic similarity.

Also, you'll do keyword matching with word vector-based semantic similarity methods. Finally, we will combine all this information to generate a semantic representation for the dataset utterances.

By the end of this chapter, you'll learn how to semantically process a real-world dataset completely. You'll learn how to extract entities, recognize intents, and perform semantic similarity calculations. The tools of this chapter are really what you'll build for a real-world **natural language processing** (**NLP**) pipeline, including an NLU chatbot and an NLU customer support application.

In this chapter, we're going to cover the following main topics:

- Extracting named entities
- Using dependency relations for intent recognition
- Semantic similarity methods for semantic parsing
- Putting it all together

Technical requirements

In this chapter, we'll process a dataset. The dataset and the chapter code can be found at `https://github.com/PacktPublishing/Mastering-spaCy/tree/main/Chapter06`.

We used the pandas library of Python to manipulate our dataset, besides using spaCy. We also used the awk command-line tool. pandas can be installed via pip and awk is preinstalled in many Linux distributions.

Extracting named entities

In many NLP applications, including semantic parsing, we start looking for meaning in a text by examining the entity types and placing an entity extraction component into our NLP pipelines. **Named entities** play a key role in understanding the meaning of user text.

We'll also start a semantic parsing pipeline by extracting the named entities from our corpus. To understand what sort of entities we want to extract, first, we'll get to know the ATIS dataset.

Getting to know the ATIS dataset

Throughout this chapter, we'll work with the ATIS corpus. ATIS is a well-known dataset; it's one of the standard benchmark datasets for intent classification. The dataset consists of customer utterances who want to book a flight, get information about the flights, including flight costs, flight destinations, and timetables.

No matter what the NLP task is, you should always go over your corpus with a naked eye. We want to get to know our corpus so that we integrate our observations of corpus into our code. While viewing our text data, we usually keep an eye on the following:

- What kind of utterances are there? Is it a short text corpus or does the corpus consist of long documents or medium-length paragraphs?

- What sort of entities does the corpus include? People's names, city names, country names, organization names, and so on. Which ones do we want to extract?

- How is punctuation used? Is the text correctly punctuated, or is no punctuation used at all?

- How are the grammatical rules followed? Is the capitalization correct? Did users follow the grammatical rules? Are there misspelled words?

Before starting any processing, we'll examine our corpus. Let's go ahead and download the dataset:

```
$ wget
https://github.com/PacktPublishing/Mastering-spaCy/blob/
main/Chapter06/data/atis_intents.csv
```

The dataset is a two-column CSV file. First, we'll get some insights into the dataset statistics with **pandas**. pandas is a popular data manipulation library that is frequently used by data scientists. You can read more at https://pandas.pydata.org/pandas-docs/version/0.15/tutorials.html:

1. Let's begin by reading the CSV file into Python. We'll use the read_csv method of pandas:

```
import pandas as pd
dataset = pd.read_csv("data/atis_intents.csv",
header=None)
```

The dataset variable holds the CSV as an object for us.

2. Next, we'll call head() on the dataset object. head() outputs the first 10 columns of the dataset:

```
dataset.head()
```

The result looks as follows:

	0	1
0	atis_flight	i want to fly from boston at 838 am and arriv...
1	atis_flight	what flights are available from pittsburgh to...
2	atis_flight_time	what is the arrival time in san francisco for...
3	atis_airfare	cheapest airfare from tacoma to orlando
4	atis_airfare	round trip fares from pittsburgh to philadelp...

Figure 6.1 – Overview of the dataset

As you see, the dataset object contains rows and columns. It is indeed a CSV object. The first column contains the intent, and the second column contains the user utterance.

3. Now we can print some example utterances:

```
for text in dataset[1].head():
    print(text)
i want to fly from boston at 838 am and arrive in denver
at 1110 in the morning
what flights are available from pittsburgh to baltimore
on thursday morning
```

```
what is the arrival time in san francisco for the 755 am
flight leaving washington
```

```
cheapest airfare from tacoma to orlando
```

```
round trip fares from pittsburgh to philadelphia under
1000 dollars
```

As we can see, the first user wants to book a flight; they included the destination, the source cities, and the flight time. The third user is asking about the arrival time of a specific flight and the fifth user made a query with a price limit. The utterances are not capitalized or punctuated. This is because these utterances are an output of a speech-to-text engine.

4. Last, we can see the distribution of the number of utterances by intent:

```
grouped = dataset.groupby(0).size()
print(grouped)
atis_abbreviation                              147
atis_aircraft                                   81
atis_aircraft#atis_flight#atis_flight_no          1
atis_airfare                                   423
atis_airfare#atis_flight_time                     1
atis_airline                                   157
atis_airline#atis_flight_no                       2
atis_airport                                    20
atis_capacity                                   16
atis_cheapest                                    1
atis_city                                       19
atis_distance                                   20
atis_flight                                   3666
```

You can find the dataset exploration code at the book's GitHub repository under https://github.com/PacktPublishing/Mastering-spaCy/blob/main/Chapter06/ATIS_dataset_exploration.ipynb.

5. After this point, we'll process just the utterance text. Hence, we can drop the first column. To do so, we'll play a small trick with the Unix tool awk:

```
awk -F ',' '{print $2}' atis_intents.csv > atis_
utterances.txt
```

Here, we printed the second column of the input CSV file (where the filed separator is a `,`) and directed the output into a text file called `atis_utterances.txt`. Now that our utterances are ready to process, we can go ahead and extract the entities.

Extracting named entities with Matcher

As we have already seen, this is a flights dataset. Hence, we expect to see city/country names, airport names, and airline names:

1. Here are some examples:

    ```
    does american airlines fly from boston to san francisco
    what flights go from dallas to tampa
    show me the flights from montreal to chicago
    what flights do you have from ontario
    The users also provide the dates, times, days of the
    weeks they wish to fly on. These entities include
    numbers, month names, day of the week names as well as
    time adverbs such as next week, today, tomorrow, next
    month. Let's see some example entities:
    list flights from atlanta to boston leaving between 6 pm
    and 10 pm on august eighth
    i need a flight after 6 pm on wednesday from oakland to
    salt lake city
    show me flights from minneapolis to seattle on july
    second
    what flights leave after 7 pm from philadelphia to boston
    ```

2. Also, the `atis_abbreviation` intent contains utterances that are inquiries about some abbreviations. Flight abbreviations can be fare codes (for example, M = Economy), airline name codes (for example, United Airlines = UA), and airport codes (for example, Berlin Airport = BER), and so on. Examples include the following:

    ```
    what does the abbreviation ua mean
    what does restriction ap 57 mean
    explain restriction ap please
    what's fare code yn
    ```

3. Let's visualize some utterances from the dataset. The following screenshot shows the entities with their types:

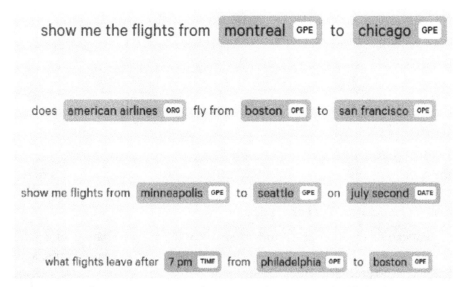

Figure 6.2 – Example corpus sentences with entities and entity types highlighted; generated by the displaCy online demo

4. We can see all the entity types and their frequencies more systematically. The following code segment makes the following actions:

a) It reads the text file of utterances that we created in the preceding dataset exploration subsection.

b) It iterates over each utterance and creates a Doc object.

c) It extracts entities from the current doc object.

d) It updates the global entity label list with the labels of the entities.

e) It finally calculates the frequency of each label with a counter object.

Here is the code:

```
from collections import Counter
import spacy
nlp = spacy.load("en_core_web_md")
corpus = open("atis_utterances.txt", "r").read().
split("\n")

all_ent_labels = []
for sentence in corpus:
        doc = nlp(sentence.strip())
        ents = doc.ents
```

```
        all_ent_labels += [ent.label_ for ent in ents]
c = Counter(all_ent_labels)
print(c)
Counter({'GPE': 9242, 'DATE': 1435, 'TIME': 872, 'ORG':
464, 'CARDINAL': 205, 'ORDINAL': 190, 'FAC': 57, 'NORP':
52, 'MONEY': 47, 'PERSON': 37, 'PRODUCT': 19, 'LOC': 4,
'QUANTITY': 3})
```

Our initial guess seems correct. The most frequent entity labels are GPE (location names), DATE, TIME, and ORGANIZATION. Obviously, the location entities refer to destination and source cities/countries, hence they play a very important role in the overall semantic success of our application.

5. We'll first extract the location entities by spaCy Matcher by searching for a pattern of the preposition location_name form. The following code extracts location entities preceded with a preposition:

```
import spacy
from spacy.matcher import Matcher
nlp = spacy.load("en_core_web_md")
matcher = Matcher(nlp.vocab)
pattern = [{"POS": "ADP"}, {"ENT_TYPE": "GPE"}]
matcher.add("prepositionLocation", [pattern])

doc = nlp("show me flights from denver to boston on
tuesday")
matches = matcher(doc)
for mid, start, end in matches:
    print(doc[start:end])
...
from denver
to boston
```

We already saw how to initialize a Matcher object and add patterns to it. Still, we'll recall how to use a Matcher object to extract the matches now. Here's what we did in this code segment:

a) We started by importing spacy and the spacy.matcher class in *lines 1-2*.

b) We created a language pipeline object, nlp, at *line 3*.

c) In *line 4*, we initialized the Matcher object with the language vocabulary.

d) In line 5, we created a pattern matching two tokens, a preposition (POS tag ADP means an *adposition = preposition + postposition*) and a location entity (label GPE means a **location entity**).

e) We added this pattern to the Matcher object.

f) Finally, we asked for the matches in an example corpus sentence and printed the matches.

6. Although the `from` and `to` prepositions dominate in this dataset, verbs about leaving and arriving can be used with a variety of prepositions. Here are some more example sentences from the dataset:

```
doc = nlp("i'm looking for a flight that goes from
ontario to westchester and stops in chicago")
matches = matcher(doc)
for mid, start, end in matches:
    print(doc[start:end])
...
from ontario
to westchester
in chicago
```

The second example sentence is a question sentence:

```
doc = nlp("what flights arrive in chicago on sunday on
continental")
matches = matcher(doc)
for mid, start, end in matches:
    print(doc[start:end])
...
in chicago
```

Another example sentence from the dataset contains an abbreviation in a destination entity:

```
doc = nlp("yes i'd like a flight from long beach to st.
louis by way of dallas")
matches = matcher(doc)
for mid, start, end in matches:
    print(doc[start:end])
...
from long
```

```
to st
of dallas
```

Our last example sentence is again a question sentence:

```
doc = nlp("what are the evening flights flying out of
dallas")
matches = matcher(doc)
for mid, start, end in matches:
    print(doc[start:end])
...
of dallas
```

Here, we see some phrasal verbs such as `arrive in`, as well as preposition and verb combinations such as `stop in` and `fly out of`. By the way of `Dallas` does not include a verb at all. The user indicated that they want to make a stop at Dallas. `to`, `from`, `in`, `out`, and `of` are common prepositions that are used in a traveling context.

7. After extracting the locations, we can now extract the airline information. The `ORG` entity label means an organization and it corresponds to airline company names in our dataset. The following code segment extracts the organization names, possibly multi-worded names:

```
pattern = [{"ENT_TYPE": "ORG", "OP": "+"}]
matcher.add("AirlineName", [pattern])
doc = nlp("what is the earliest united airlines flight
flying from denver")
matches = matcher(doc)
for mid,start,end in matches:
    print(doc[start:end])
...
united
united airlines
airlines
```

Here, we extracted the entities whose labels are `ORG`. We wanted to capture one or more occurrences to capture the multi-word entities as well, which is why we used the `OP: "+"` operator.

8. Extracting dates and times are not very different; you can replicate the preceding with code with `ENT_TYPE: DATE` and `ENT_TYPE: TIME`. We encourage you to try it yourself. The following screenshot exhibits how the date and time entities look in detail:

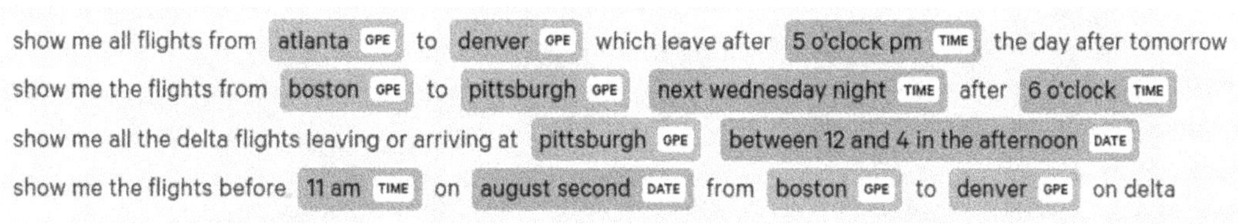

show me all flights from atlanta GPE to denver GPE which leave after 5 o'clock pm TIME the day after tomorrow

show me the flights from boston GPE to pittsburgh GPE   next wednesday night TIME after 6 o'clock TIME

show me all the delta flights leaving or arriving at pittsburgh GPE   between 12 and 4 in the afternoon DATE

show me the flights before 11 am TIME on august second DATE from boston GPE to denver GPE on delta

Figure 6.3 – Example dataset sentences with date and time entities highlighted. The image is generated by the displaCy online demo

9. Next, we'll extract abbreviation type entities. Extracting the abbreviation entities is a bit trickier. First, we will have a look at how the abbreviations appear:

```
what does restriction ap 57 mean?
what does the abbreviation co mean?
what does fare code qo mean
what is the abbreviation d10
what does code y mean
what does the fare code f and fn mean
what is booking class c
```

Only one of these sentences includes an entity. The first example sentence includes an AMOUNT entity, which is 57. Other than that, abbreviations are not marked with any entity type at all. In this case, we have to provide some custom rules to the Matcher. Let's make some observations first, and then form a Matcher pattern:

a) An abbreviation can be broken into two parts – letters, and digits.

b) The letter part can be 1-2 characters long.

c) The digit part is also 1-2 characters long.

d) The presence of digits indicates an abbreviation entity.

e) The presence of the following words indicates an abbreviation entity: class, code, abbreviation.

f) The POS tag of an abbreviation is a noun. If the candidate word is a 1-letter or 2-letter word, then we can look at the POS tag and see whether it's a noun. This approach eliminates the false positives, such as *us* (pronoun), *me* (pronoun), *a* (determiner), and *an* (determiner).

10. Let's now put these observations into Matcher patterns:

```
pattern1 = [{"TEXT": {"REGEX": "\w{1,2}\d{1,2}"}}]
pattern2 = [{"SHAPE": { "IN": ["x", "xx"]}}, {"SHAPE": {
"IN": ["d", "dd"]}}]
pattern3 = [{"TEXT": {"IN": ["class", "code", "abbrev",
"abbreviation"]}}, {"SHAPE": { "IN": ["x", "xx"]}}]
pattern4 =    [{"POS": "NOUN", "SHAPE": { "IN": ["x",
"xx"]}}]
```

Then we create a Matcher object with the patterns we defined:

```
matcher = Matcher(nlp.vocab)
matcher.add("abbrevEntities", [pattern1, pattern2,
pattern3, pattern4])
```

We're now ready to feed our sentences into the matcher:

```
sentences = [
'what does restriction ap 57 mean',
'what does the abbreviation co mean',
'what does fare code qo mean',
  'what is the abbreviation d10',
'what does code y mean',
'what does the fare code f and fn mean',
  'what is booking class c'
]
        18. We're ready to feed our sentences to the
matcher:for sent in sentences:
    doc = nlp(sent)
    matches = matcher(doc)
    for mid, start, end in matches:
      print(doc[start:end])
...
ap 57
57
abbreviation co
co
code qo
d10
code y
```

```
code f
class c
c
```

In the preceding code, we defined four patterns:

a) The first pattern matches to a single token, which consists of 1-2 letters and 1-2 digits. For example, `d1`, `d10`, `ad1`, and `ad21` will match this pattern.

b) The second pattern matches to 2-token abbreviations where the first token is 1-2 letters and the second token 1-2 digits. The abbreviations `ap 5`, `ap 57`, `a 5`, and `a 57` will match this pattern.

c) The third pattern matches to two tokens too. The first token is a context clue word, such as `class` or `code`, and the second token should be a 1-2 letter token. Some example matches are `code f`, `code y`, and `class c`.

d) The fourth pattern extracts 1-2 letter short words whose POS tag is NOUN. Some example matches from the preceding sentences are `c` and `co`.

spaCy Matcher makes life easy for us by allowing us to make use of token shape, context clues, and a token POS tag. We made a very successful entity extraction in this subsection by extracting locations, airline names, dates, times, and abbreviations. In the next subsection, we'll go deeper into the sentence syntax and extract entities from the sentences where context does not offer many clues.

Using dependency trees for extracting entities

In the previous subsection, we extracted entities where the context provides obvious clues. Extracting the destination city from the following sentence is easy. We can look for the `to` + GPE pattern:

```
I want to fly to Munich tomorrow.
```

But suppose the user provides one of the following sentences instead:

```
I'm going to a conference in Munich. I need an air ticket.
My sister's wedding will be held in Munich. I'd like to book a
flight.
```

Here, the preposition to refers to conference, not Munich, in the first sentence. In this sentence, we need a pattern such as to + ... + GPE. Then, we have to be careful what words can come in between "to" and the city name, as well as what words should not come. For instance, this sentence carries a completely different meaning and shouldn't match:

```
I want to book a flight to my conference without stopping at
Berlin.
```

In the second sentence, there's no to at all. Here, as we see from these examples, we need to examine the syntactic relations between words. In *Chapter 3, Linguistic Features*, we already saw how to interpret dependency trees to understand the relations between words. In this subsection, we'll walk the dependency trees.

Walking a dependency tree means visiting the tokens in a custom order, not necessarily from left to right. Usually, we stop iterating over the dependency tree once we find what we're looking for. Again, a dependency tree shows the syntactic relations between its words. Recall from *Chapter 3, Linguistic Features*, that the relations are represented with directed arrows, connecting the head and the child of a relation. Every word in a sentence has to involve at least one relation. This fact guarantees that we'll visit each word while walking through the sentence.

> **Recall**
>
> Before proceeding to code, first, let's remember some concepts about dependency trees. ROOT is a special dependency label and is always assigned to the main verb of the sentence. spaCy shows syntactic relations with arcs. One of the tokens is the syntactic parent (called the HEAD) and the other is dependent (called the CHILD). By way of an example, in *Figure 6.3*, **going** has 3 syntactic children – **I**, **m**, and **to**. Equivalently, the syntactic head of **to** is **going** (the same applies to **I** and **m**).

Coming back to our examples, we'll iterate the utterance dependency trees to find out whether the preposition **to** is syntactically related to the location entity, **Munich**. First of all, let's see the dependency parse of our example sentence *I'm going to a conference in Munich* and also remember what a dependency tree looks like:

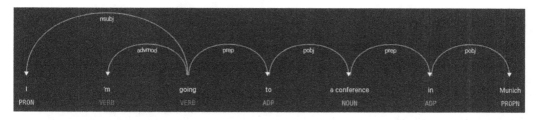

Figure 6.4 – Dependency parse of the example sentence

There are no incoming arcs into the verb **going**, so **going** is the root of the dependency tree (when we examine the code, we'll see that the dependency label is **ROOT**). This is supposed to happen because **going** is the main verb of the sentence. If we follow the arc to its immediate right, we encounter **to**; jumping over the arcs to the right we reach **Munich**. This shows that there's a syntactic relation between **to** and **Munich**.

Let's now iterate over the dependency tree with code. There are two possible ways to connect **to** and **Munich**:

- Left to right. We start from **to** and try to reach **Munich** by visiting "to"'s syntactic children. This approach may not be a very good idea, because if "to" has more than one child, then we need to check each child and keep track of all the possible paths.

- Right to left. We start from **Munich**, jump onto its head, and follow the head's head, and so on. Since each word has exactly one head, it's guaranteed that there will be only one path. Then we determine whether **to** is on this path or not.

The following code segments implement the second approach, start the dependency tree walk from Munich, and look for to:

```python
import spacy
nlp = spacy.load("en_core_web_md")


def reach_parent(source_token, dest_token):
    source_token = source_token.head
    while source_token != dest_token:
        if source_token.head == source_token:
            return None
        source_token = source_token.head
    return source_token


doc = nlp("I'm going to a conference in Munich.")
doc[-2]
```

```
Munich
doc[3]
to
doc[-1]
.
reach_parent(doc[-2], doc[3])
to
reach_parent(doc[-1], doc[3])
None
```

In the `reach_parent` function, the following applies:

- We start from a source token and try to reach the destination token.

- In the `while` loop, we iterate over the head of each token, starting from the source token.

- The loop stops when we reach either the source token or the root of the sentence.

- We test for reaching the root via the `source_token == source_token.head` line. Because the root token always refers to itself, its head is itself (remember that in the dependency tree, there are no incoming arcs into the root).

- Finally, we tested our function on two different test cases. In the first one, the source and destination are related, whereas in the second test, there's no relation, hence the function returns `None`.

This approach is very different from the previous subsection's rather straightforward approach. Natural language is complicated and challenging to process, and it's important to know the approaches available and have the necessary tools in your toolbox when you need to use them.

Next, we'll dive into a popular subject – intent recognition with a syntactic approach. Let's see some key points of designing a good intent recognition component, including recognizing multiple intents as well.

Using dependency relations for intent recognition

After extracting the entities, we want to find out what sort of intent the user carries – to book a flight, to purchase a meal on their already booked flight, cancel their flight, and so on. If you look at the intents list again, you will see that every intent includes a verb (to book) and an object that the verb acts on (flight, hotel, meal).

In this section, we'll extract transitive verbs and their direct objects from utterances. We'll begin our intent recognition section by extracting the transitive verb and the direct object of the verb. Then, we'll explore how to understand a user's intent by recognizing synonyms of verbs and nouns. Finally, we'll see how to determine a user's intent with semantic similarity methods. Before we move on to extracting transitive verbs and their direct objects, let's first quickly go over the concepts of transitive verbs and direct/indirect objects.

Linguistic primer

In this section, we'll explore some linguistic concepts related to sentence structure, including verbs and verb-object relations. A verb is a very important component of the sentence as it indicates the action in the sentence. The object of the sentence is the thing/person that is affected by the action of the verb. Hence, there's a natural connection between the sentence verb and objects. The concept of transitivity captures verb-object relations. A transitive verb is a verb that needs an object to act upon. Let's see some examples:

```
I bought flowers.
He loved his cat.
He borrowed my book.
```

In these example sentences, bought, loved, and borrowed are transitive verbs. In the first sentence, bought is the transitive verb and flowers is its object, the thing that has been bought by the sentence subject, I. Loved – his cat and borrowed – my book are transitive verb-object examples. We'll focus on the first sentence again - what happens if we erase the flowers object?

```
I bought
```

Bought **what**? Without an object, this sentence doesn't carry any meaning at all. In the preceding sentences, each of the objects completes the meaning of the verb. This is a way of understanding whether a verb is transitive or not – erase the object and check whether the sentence remains semantically intact.

Some verbs are transitive and some verbs are intransitive. An **intransitive verb** is the opposite of a transitive verb; it doesn't need an object to act upon. Let's see some examples:

```
Yesterday I slept for 8 hours.
The cat ran towards me.
When I went out, the sun was shining.
Her cat died 3 days ago.
```

In all the preceding sentences, the verbs make sense without an object. If we erase all the words other than the subject and object, these sentences are still meaningful:

```
I slept.
The cat ran.
The sun was shining.
Her cat died.
```

Pairing an intransitive verb with an object doesn't make sense. You can't run someone/something, you can't shine something/someone, and you certainly cannot die something/someone.

Sentence object

As we remarked before, the object is the thing/person that is affected by the verb's action. The action stated by the verb is committed by the sentence subject and the sentence object gets affected.

A sentence can be direct or indirect. A **direct object** answers the questions **whom? / what?** You can find the direct object by asking **The subject {verb} what/who?**. Here are some examples:

```
I bought flowers.  I bought what? - flowers
He loved his cat.  He loved who?  - his cat
He borrowed my book. He borrowed what? - my book
```

An **indirect object** answers the questions **for what?/for whom?/to whom?**. Let's see some examples:

```
He gave me his book.  He gave his book to whom?  - me
He gave his book to me. He gave his book to whom? -me
```

Indirect objects are often preceded by the prepositions to, for, from, and so on. As you can see from these examples, an indirect object is also an object and is affected by the verb's action, but its role in the sentence is a bit different. An indirect object is sometimes viewed as the recipient of the direct object.

This is all you need to know about transitive/intransitive verbs and direct/indirect objects to digest this chapter's material. If you want to learn more about sentence syntax, you can read the great book *Linguistic Fundamentals for Natural Language Processing* by **Emily Bender**: (`https://dl.acm.org/doi/book/10.5555/2534456`). We have covered the basics of sentence syntax, but this is still a great resource to learn about syntax in depth.

Extracting transitive verbs and their direct objects

While recognizing the intent, usually we apply these steps to a user utterance:

1. Splitting the sentence into tokens.

2. Dependency parsing is performed by spaCy. We walk the dependency tree to extract the tokens and relations that we're interested in, which are the verb and the direct object, as shown in the following diagram:

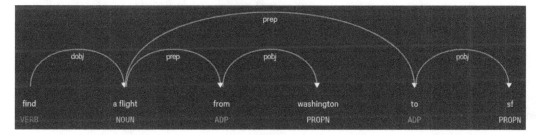

Figure 6.5 – Dependency parse of an example sentence from the corpus

In this example sentence, the transitive verb is **find** and the direct object is **a flight**. The relation **dobj** connects a transitive verb to its direct object. If we follow the arc, semantically, we see that the user wants to commit the action of finding and the object they want to find is a flight. We can merge **find** and **a flight** into a single word, **findAflight** or **findFlight**, which can be this intent's name. Other intents can be **bookFlight**, **cancelFlight**, **bookMeal**, and so on.

Let's extract the verb and the direct object in a more systematic way. We'll first spot the direct object by looking for the dobj label in the sentence. To locate the transitive verb, we look at the direct object's syntactic head. A sentence can include more than one verb, hence we're careful while processing the verbs. Here is the code:

```
import spacy
nlp = spacy.load("en_core_web_md")
doc = nlp("find a flight from washington to sf")
for token in doc:
  if token.dep_ == "dobj":
    print(token.head.text + token.text.capitalize())
findFlight
```

In this code segment, the following applies:

- We applied the pipeline to our sample sentence.
- Next, we spotted the direct object by looking for a token whose dependency label is dobj.
- When we located a direct object, we spotted the corresponding transitive verb by obtaining the direct object's syntactic head.
- Finally, we printed the verb and the object to generate this intent's name.

Great! Intent recognition was successful! Here, we recognized a single intent. Some utterances may carry multiple intents. In the next section, we'll learn how to recognize multiple intents based on the techniques of this section.

Extracting multiple intents with conjunction relation

Some utterances carry multiple intents. For example, consider the following utterance from the corpus:

```
show all flights and fares from denver to san francisco
```

Here, the user wants to list all the flights and, at the same time, wants to see the fare info. One way of processing is considering these intents as a single and complex intent. In that case, we can express this complex intent as action: show, objects: flights, fares. Another and more common way of processing this sort of utterance is to label the utterance with multiple intents. In the dataset, this example utterance is marked with two intents as atis_flight#atis_airfare:

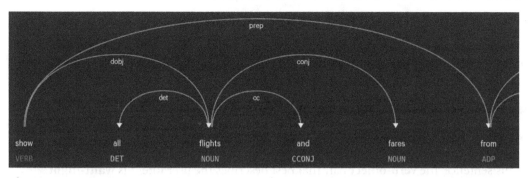

Figure 6.6 – Dependency tree of the example dataset utterance. The conj relation is between "flight" and "fares"

In the preceding diagram, we see that the **dobj** arc connects **show** and **flights**. The **conj** arc connects **flights** and **fares** to indicate the conjunction relation between them. The conjunction relation is built by a conjunction such as **and** or **or** and indicates that a noun is joined to another noun by this conjunction. In this situation, we extract the direct object and its conjuncts. Let's now see how we can turn this process into code:

```
import spacy
nlp = spacy.load("en_core_web_md")

doc = nlp("show all flights and fares from denver to san
francisco")
for token in doc:
    if token.dep_ == "dobj":
        dobj = token.text
        conj = [t.text for t in token.conjuncts]
        verb = donj.head
print(verb, dobj, conj)
show flights ['fares']
```

Here, we looped over all tokens to locate the direct object of the sentence. When we found the direct object, we obtained its conjuncts. After that, finding the transitive verb is the same as we did in the previous code segment, we extracted the direct object's head. After extracting the verb and two objects, if we want, we can combine the two to create two intent names – showFlights and showFares.

Recognizing the intent using wordlists

In some cases, tokens other than the transitive verb and the direct object contain the semantics of the user intent. In that case, you need to go further down in the syntactic relations and explore the sentence structure deeper.

As an example, consider the following utterance from our dataset:

```
i want to make a reservation for a flight
```

In this sentence, the **verb-object** pair that best describes the user intent is **want-flight**. However, if we look at the parse tree in *Figure 6.7*, we see that **want** and **flight** are not directly related in the parse tree. **want** is related to the transitive verb **make**, and **flight** is related to the direct object **reservation**, respectively:

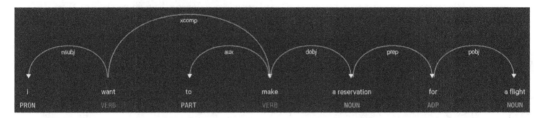

Figure 6.7 – Parse tree of the example sentence from the dataset

What will we do then? We can play a trick and keep a list of helper verbs such as **would like**, **want**, **make**, and **need**. Here's the code:

```
doc = nlp("i want to make a reservation for a flight")
dObj =None
tVerb = None
# Extract the direct object and its transitive verb
for token in doc:
    If token.dep_ == "dobj":
        dObj = token
        tVerb = token.head
# Extract the helper verb
intentVerb = None
verbList = ["want", "like", "need", "order"]
if tVerb.text in verbList:
    intentVerb = tVerb
else:
    if tVerb.head.dep_ == "ROOT":
```

```
        helperVerb = tVerb.head
# Extract the object of the intent
intentObj = None
objList = ["flight", "meal", "booking"]
if dObj.text in objList:
    intentObj = dObj
else:
    for child in dObj.children:
        if child.dep_ == "prep":
            intentObj = list(child.children)[0]
            break
        elif child.dep_ == "compound":
            intentObj = child
            break
print(intentVerb.text + intentObj.text.capitalize())
wantFlight
```

Here's what we did step by step:

1. We started by locating the direct object and its transitive verb.

2. Once we found them, we compared them against our predefined lists of words. For this example, we used two shortened lists, `verbList` contains a list of helper verbs, and `objList` contains a list of the possible object words we want to extract.

3. We checked the transitive verb. If it's not in the list of helper verbs, then we checked the main verb of the sentence (marked by ROOT), which is the head of the transitive verb. If the transitive verb is the main verb of the sentence, then the syntactic head of this verb is itself (`tVerb.head` is `tVerb`). Hence, the line `if tVerb.head.dep_ == "ROOT"` evaluates to `True` and this implementation works.

4. Next, we checked the direct object. If it's not in the list of possible objects, then we check its syntactic children. For each child, we check whether the child is a preposition of the direct object. If so, we pick up the child's child (it can have only one child).

5. Finally, we printed the string that represents the intent name, which is `wantFlight`.

At this point, take a deep breath. It takes time to digest and process information, especially when it's about sentence syntax. You can try different sentences from the corpus and see what the script does by checkpointing/putting prints into the code.

In the next section, we'll explore a very handy tool, using synonym lists. Let's move ahead to the next section and learn how to make the best of semantic similarity.

Semantic similarity methods for semantic parsing

Natural language allows us to express the same concept in different ways and with different words. Every language has synonyms and semantically related words.

As an NLP developer, while developing a semantic parser for a chatbot application, text classification, or any other semantic application, you should keep in my mind that users use a fairly wide set of phrases and expressions for each intent. In fact, if you're building a chatbot by using a platform such as RASA (`https://rasa.com/`) or on a platform such as Dialogflow (`https://dialogflow.cloud.google.com/`), you're asked to provide as many utterance examples as you can provide for each intent. Then, these utterances are used to train the intent classifier behind the scenes.

There are usually two ways to recognize semantic similarity, either with a synonyms dictionary or with word vector-based semantic similarity methods. In this section, we will discuss both approaches. Let's start with how to use a synonyms dictionary to detect semantic similarity.

Using synonyms lists for semantic similarity

We already went through our dataset and saw that different verbs are used to express the same actions. For instance, **landing**, **arriving**, and **flying to** verbs carry the same meaning, whereas **leaving**, **departing**, and **flying from** verbs form another semantic group.

We already saw that in most cases, the transitive verbs and direct objects express the intent. An easy way to determine whether two utterances represent the same intent is to check whether the verbs and the direct objects are synonyms.

Let's take an example and compare two example utterances from the dataset. First, we prepare a small synonyms dictionary. We include only the base forms of the verbs and nouns. While doing the comparison, we also use the base form of the words:

```
verbSynsets = [
("show", "list"),
```

```
("book", "make a reservation", "buy", "reserve")
]
objSynsets = [
("meal", "food"),
("aircraft", "airplane", "plane")
]
```

Each **synonym set (synset)** includes the set of synonyms for our domain. We usually include the language-general synonyms (airplane-plane) and the domain-specific synonyms (book-buy).

The synsets are ready to use, and we're ready to move onto the spaCy code. Let's go step by step:

1. First, we construct two doc objects corresponding to the two utterances we want to compare, doc and doc2:

   ```
   doc = nlp("show me all aircrafts that cp uses")
   doc2 = nlp("list all meals on my flight")
   ```

2. Then, we extract the transitive verb and direct object of the first utterance:

   ```
   for token in doc:
       if token.dep_ == "dobj":
           obj = token.lemma_
           verb = token.head.lemma_
           break
   ```

3. Then we do the same for the second utterance:

   ```
   for token in doc2:
       if token.dep_ == "dobj":
           obj2 = token.lemma_
           verb2 = token.head.lemma_
           break

   verb, obj
   ('show' , 'aircraft')
   verb2, obj2
   ('list', 'meal')
   ```

4.   We obtained a synset of the first verb shown. Then, we checked whether the second verb list is in this synset, which returns `True`:

```
vsyn = [syn for syn in verbSynsets if verb in item]
vsyn[0]
("show", "list")
v2 in vsyn[0]
True
```

5.   Similarly, we obtain the synset of the first direct object – `aircraft`. Then we check whether the second direct object `meal` is in this synset, which is obviously not true:

```
osyn = [syn for syn in objSynsets if obj in item]
osyn[0]
("aircraft", "airplane", "plane")
obj2 in vsyn[0]
False
```

6.   We deduce that the preceding two utterances do not refer to the same intent.

Synonym lists are great for semantic similarity calculations, and many real-world NLP applications benefit from such precompiled lists. Using synonyms is not always applicable though. Making a dictionary look up each word in a sentence can become inefficient for big synsets. In the next section, we'll introduce a more efficient way of calculating semantic similarity with word vectors.

Using word vectors to recognize semantic similarity

In *Chapter 5, Working with Word Vectors and Semantic Similarity*, we already saw that the word vector carries semantics, including synonymity information. Synonym lists are handy if you work in a very specific domain and the number of synonyms is rather low. Working with big synsets can become inefficient at some point because we have to make a dictionary look up the verbs and direct objects each time. However, word vectors offer us a very convenient and vector-based way to calculate semantic similarity.

Let's go over the code from the previous subsection again. This time, we'll calculate the semantic distance between words with spaCy word vectors. Let's go step by step:

1.   First, we construct two `doc` objects that we want to compare:

```
doc = nlp("show me all aircrafts that cp uses")
doc2 = nlp("list all meals on my flight")
```

2. Then we extract the verb and object of the first sentence:

```
for token in doc:
    if token.dep_ == "dobj":
        obj = token
        verb = token.head
        break
```

3. We repeat the same procedure on the second sentence:

```
for token in doc2:
    if token.dep_ == "dobj":
        obj2 = token
        verb2 = token.head
        break
verb, obj
('show' , 'aircraft')
verb2, obj2
('list', 'meal')
```

4. Now, we calculate the semantic similarity between two direct objects using the word vector-based similarity method of spaCy:

```
obj.similarity(obj2)
0.15025872                    # A very low score, we can
deduce these 2 utterances are not related at this point.
```

5. Finally, we calculate the similarity between the verbs:

```
verb.similarity(verb2)
0.33161193
```

The preceding code is different from the previous code. This time, we used the token objects directly; no lemma extraction is required. Then we called the `token.similarity(token2)` method of spaCy to calculate the semantic distance between the direct objects. The resulting score is very low. At this point, we deduce that these two utterances do not represent the same intent.

This is an easy and efficient way of calculating semantic similarity. We remarked in the very first chapter that spaCy provides easy-to-use and efficient tools for NLP developers, and now we can see why.

Putting it all together

We already extracted the entities and recognized the intent in several ways. We're now ready to put it all together to calculate a semantic representation for a user utterance!

1. We'll process the example dataset utterance:

```
show me flights from denver to philadelphia on tuesday
```

We'll hold a dictionary object to hold the result. The result will include the entities and the intent.

2. Let's extract the entities:

```
import spacy
from spacy.matcher import Matcher
nlp = spacy.load("en_core_web_md")
matcher = Matcher(nlp.vocab)
pattern = [{"POS": "ADP"}, {"ENT_TYPE": "GPE"}]
matcher.add("prepositionLocation", [pattern])


# Location entities
doc = nlp("show me flights from denver to philadelphia on
tuesday")
matches = matcher(doc)
for mid, start, end in matches:
    print(doc[start:end])
...
from denver
to philadelphia


# All entities:
ents = doc.ents
(denver, philedelphia, tuesday)
```

3. With this information, we can generate the following semantic representation:

```
{
'utterance': 'show me flights from denver to philadelphia
on tuesday',
'entities': {
```

```
                          'date': 'tuesday',
                          'locations': {
                                           'from': 'denver',
                                           'to': 'philadelphia'
                                           }
                          }
          }
```

4. Next, we'll perform intent recognition to generate a complete semantic parsing:

```
import spacy
nlp = spacy.load("en_core_web_md")
doc = nlp("show me flights from denver to philadelphia on
tuesday")
for token in doc:
  if token.dep_ == "dobj":
    print(token.head.lemma_ + token.lemma_.capitalize())
showFlight
```

5. After determining the intent, our semantic parse for this utterance now looks like this:

```
{
'utterance': 'show me flights from denver to philadelphia
on tuesday',
'intent ': ' showFlight',
'entities': {
                 'date': 'tuesday',
                 'locations': {
                                  'from': 'denver',
                                  'to': 'philadelphia'
                                  }
                 }
}
```

The final result is that the complete semantic representation of this utterance, intent, and entities is extracted. This is a machine-readable and usable output. We pass this result to the system component that made the call to the NLP application to generate a response action.

Summary

Congratulations! You have made it to the end of a very intense chapter!

In this chapter, you learned how to generate a complete semantic parse of utterances. First, you made a discovery on your dataset to get insights about the dataset analytics. Then, you learned to extract entities with two different techniques – with spaCy Matcher and by walking on the dependency tree. Next, you learned different ways of performing intent recognition by analyzing the sentence structure. Finally, you put all the information together to generate a semantic parse.

In the next chapters, we will shift toward more machine learning methods. The next section concerns how to train spaCy NLP pipeline components on your own data. Let's move ahead and customize spaCy for ourselves!

Section 3: Machine Learning with spaCy

This section talks about how building advanced NLP models with spaCy takes knowledge, analysis, and practice. You will experiment with different NLP machine learning tasks while customizing your own statistical models, experimenting with brand-new transformers, and designing your own NLP pipelines. You will learn tips and tricks along the way from an NLP master.

This section comprises the following chapters:

- *Chapter 7, Customizing spaCy Models*
- *Chapter 8, Text Classification with spaCy*
- *Chapter 9, spaCy and Transformers*
- *Chapter 10, Putting Everything Together – Designing Your Chatbot with spaCy*

7
Customizing spaCy Models

In this chapter, you will learn how to train, store, and use custom statistical pipeline components. First, we will discuss when exactly we should perform custom model training. Then, you will learn a fundamental step of model training – how to collect and label your own data.

In this chapter, you will also learn how to make the best use of **Prodigy**, the annotation tool. Next, you will learn how to update an existing statistical pipeline component with your own data. We will update the spaCy pipeline's **named entity recognizer (NER)** component with our own labeled data.

Finally, you will learn how to create a statistical pipeline component from scratch with your own data and labels. For this purpose, we will again train an NER model. This chapter takes you through a complete machine learning practice, including collecting data, annotating data, and training a model for information extraction.

By the end of this chapter, you'll be ready to train spaCy models on your own data. You'll have the full skillset of collecting data, preprocessing data in to the format that spaCy can recognize, and finally, training spaCy models with this data. In this chapter, we're going to cover the following main topics:

- Getting started with data preparation

- Annotating and preparing data

- Updating an existing pipeline component

- Training a pipeline component from scratch

Technical requirements

The chapter code can be found at the book's GitHub repository: `https://github.com/PacktPublishing/Mastering-spaCy/tree/main/Chapter07`.

Getting started with data preparation

In the previous chapters, we saw how to make the best of spaCy's pre-trained statistical models (including the **POS tagger**, NER, and **dependency parser**) in our applications. In this chapter, we will see how to customize the statistical models for our custom domain and data.

spaCy models are very successful for general NLP purposes, such as understanding a sentence's syntax, splitting a paragraph into sentences, and extracting some entities. However, sometimes, we work on very specific domains that spaCy models didn't see during training.

For example, the Twitter text contains many non-regular words, such as hashtags, emoticons, and mentions. Also, Twitter sentences are usually just phrases, not full sentences. Here, it's entirely reasonable that spaCy's POS tagger performs in a substandard manner as the POS tagger is trained on full, grammatically correct English sentences.

Another example is the medical domain. The medical domain contains many entities, such as drug, disease, and chemical compound names. These entities are not expected to be recognized by spaCy's NER model because it has no disease or drug entity labels. NER does not know anything about the medical domain at all.

Training your custom models requires time and effort. Before even starting the training process, you should decide *whether the training is really necessary*. To determine whether you really need custom training, you will need to ask yourself the following questions:

- Do spaCy models perform well enough on your data?
- Does your domain include many labels that are absent in spaCy models?
- Is there a pre-trained model/application in GitHub or elsewhere already? (We wouldn't want to reinvent the wheel.)

Let's discuss these questions in detail in the following sections.

Do spaCy models perform well enough on your data?

If the model performs well enough (above 0.75 accuracy), then you can customize the model output by means of another spaCy component. For example, let's say we work on the navigation domain and we have utterances such as the following:

```
navigate to my home
navigate to Oxford Street
```

Let's see what entities spaCy's NER model outputs for these sentences:

```
import spacy
nlp = spacy.load("en_core_web_md")
doc1 = nlp("navigate to my home")
doc1.ents
()
doc2 = nlp("navigate to Oxford Street")
doc2.ents
(Oxford Street,)
doc2.ents[0].label_
'FAC'
spacy.explain("FAC")
'Buildings, airports, highways, bridges, etc.'
```

Here, home isn't recognized as an entity at all, but we want it to be recognized as a location entity. Also, spaCy's NER model labels Oxford Street as FAC, which means a building/highway/airport/bridge type entity, which is not what we want.

We want this entity to be recognized as GPE, a location. Here, we can train NER further to recognize street names as GPE, as well as also recognizing some location words, such as *work*, *home*, and *my mama's house*, as GPE.

Another example is the newspaper domain. In this domain, person, place, date, time, and organization entities are extracted, but you need one more entity type – vehicle (car, bus, airplane, and so on). Hence, instead of training from scratch, you can add a new entity type by using spaCy's EntityRuler (explained in *Chapter 4, Rule-Based Matching*). Always examine your data first and calculate the spaCy models' success rate. If the success rate is satisfying, then use other spaCy components to customize.

Does your domain include many labels that are absent in spaCy models?

For instance, in the preceding newspaper example, only one entity label, vehicle, is missing from the spaCy's NER model's labels. Other entity types are recognized. In this case, you don't need custom training.

Consider the medical domain again. The entities are diseases, symptoms, drugs, dosages, chemical compound names, and so on. This is a specialized and long list of entities. Obviously, for the medical domain, you require custom model training.

If we need custom model training, we usually follow these steps:

1. Collect your data.
2. Annotate your data.
3. Decide to update an existing model or train a model from scratch.

In the data collection step, we decide how much data to collect: 1,000 sentences, 5,000 sentences, or more. The amount of data depends on the complexity of your task and domain. Usually, we start with an acceptable amount of data, make a first model training, and see how it performs; then we can add more data and retrain the model.

After collecting your dataset, you need to annotate your data in such a way that the spaCy training code recognizes it. In the next section, we will see the training data format and how to annotate data with spaCy's Prodigy tool.

The third point is to decide on training a blank model from scratch or make updates to an existing model. Here, the rule of thumb is as follows: if your entities/labels are present in the existing model but you don't see a very good performance, then update the model with your own data, such as in the preceding navigation example. If your entities are not present in the current spaCy model at all, then most probably you need custom training.

> **Tip**
> Don't rush into training your own models. First, examine if you really need
> to customize the models. Always keep in mind that training a model from
> scratch requires data preparation, training a model, and saving it, which means
> spending your time, money, and effort. Good engineering is about spending
> your resources wisely.

We'll start our journey of building a model with the first step: preparing our training data.
Let's move on to the next section and see how to prepare and annotate our training data.

Annotating and preparing data

The first step of training a model is always preparing training data. You usually collect data
from customer logs and then turn them into a dataset by dumping the data as a CSV file
or a JSON file. spaCy model training code works with JSON files, so we will be working
with JSON files in this chapter.

After collecting our data, we **annotate** our data. Annotation means labeling the intent,
entities, POS tags, and so on.

This is an example of annotated data:

```
{
"sentence": "I visited JFK Airport."
"entities": {
            "label": "LOC"
            "value": "JFK Airport"
}
```

As you see, we point the statistical algorithm to *what we want the model to learn*. In this
example, we want the model to learn about the entities, hence, we feed examples with
entities annotated.

Writing down JSON files manually can be error-prone and time-consuming. Hence, in
this section, we'll also see spaCy's annotation tool, Prodigy, along with an open source
data annotation tool, **Brat**. Prodigy is not open source or free, but we will go over how
it works to give you a better view of how annotation tools work in general. Brat is open
source and immediately available for your use.

Annotating data with Prodigy

Prodigy is a modern tool for data annotation. We will be using the Prodigy web demo (`https://prodi.gy/demo`) to exhibit how an annotation tool works.

Let's get started:

1. We navigate to the Prodigy web demo and view an example text by Prodigy, to be annotated as seen in the following screenshot:

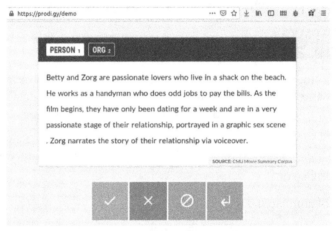

Figure 7.1 – Prodigy interface; photo taken from their web demo page

The preceding screenshot shows an example text that we want to annotate. The buttons at the bottom of the screenshot showcase the means to accept this training example, to reject this example, or to ignore this example. If the example is irrelevant to our domain/task (but involved in the dataset somehow), we ignore this example. If the text is relevant and the annotation is good, then we accept this example, and it joins our dataset.

2. Next, we'll label the entities. Labeling an entity is easy. First, we select an entity type from the upper bar (here, this corpus includes two types of entities, PERSON and ORG. Which entities you want to annotate depends on you; these are the labels you provide to the tool.) Then, we'll just select the words we want to label as an entity with the cursor, as seen in the following screenshot:

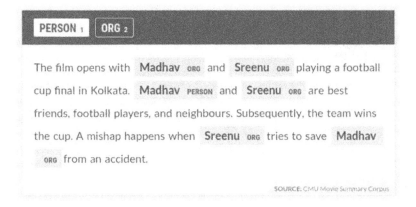

Figure 7.2 – Annotating PERSON entities on the web demo

After we're finished with annotating the text, we click the accept button. Once the session is finished, you can dump the annotated data as a JSON file. When you're finished with your annotation job, you can click the **Save** button to finish the session properly. Clicking **Save** will dump the annotated data as a JSON file automatically. That's it. Prodigy offers a really efficient way of annotating your data.

Annotating data with Brat

Another annotation tool is **Brat**, which is a free and web-based tool for text annotation (`https://brat.nlplab.org/introduction.html`). It's possible to annotate relations as well as entities in Brat. You can also download Brat onto your local machine and use it for annotation tasks. Basically, you upload your dataset to Brat and annotate the text on the interface. The following screenshot shows an annotated sentence from an example of a CoNLL dataset:

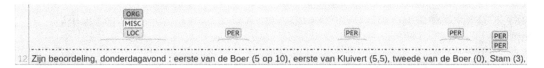

Figure 7.3 – An example annotated sentence

You can play with example datasets on the Brat demo website (`https://brat.nlplab.org/examples.html`) or get started by uploading a small subset of your own data. After the annotation session is finished, Brat dumps a JSON of annotated data as well.

spaCy training data format

As we remarked earlier, spaCy training code works with JSON file format. Let's see the details of training the data format.

For the NER, you need to provide a list of pairs of sentences and their annotations. Each annotation should include the entity type, the start position of the entity in terms of characters, and the end position of the entity in terms of characters. Let's see an example of a dataset:

```
training_data = [
("I will visit you in Munich.",  {"entities": [(20, 26,
"GPE")]}),
("I'm going to Victoria's house.", {

"entities": [

(13, 23, "PERSON"),

(24, 29, "GPE")
                                                       ]})
("I go there.", {"entities": []})
]
```

This dataset consists of three example pairs. Each example pair includes a sentence as the first element. The second element of the pair is a list of annotated entities. In the first example sentence, there is only one entity, Munich. This entity's label is GPE and starts at the 20th character position in the sentence and ends at the 25th character. Similarly, the second sentence includes two entities; one is PERSON, Victoria's, and the second entity is GPE, house. The third sentence does not include any entities, hence the list is empty.

We cannot feed the raw text and annotations directly to spaCy. Instead, we need to create an Example object for each training example. Let's see the code:

```
import spacy
from spacy.training import Example
nlp = spacy.load("en_core_web_md")
doc = nlp("I will visit you in Munich.")
annotations =  {"entities": [(20, 26, "GPE")]}
example_sent = Example.from_dict(doc, annotations)
```

In this code segment, first, we created a doc object from the example sentence. Then we fed the doc object and its annotations in a dictionary form to create an `Example` object. We'll use `Example` objects in the next section's training code.

Creating example sentences for training the dependency parser is a bit different, and we'll cover this in the *Training a pipeline component from scratch* section.

Now, we're ready to train our own spaCy models. We'll first see how to update an NLP pipeline statistical model. For this purpose, we'll train the NER component further with the help of our own examples.

Updating an existing pipeline component

In this section, we will train spaCy's NER component further with our own examples to recognize the navigation domain. We already saw some examples of navigation domain utterances and how spaCy's NER model labeled entities of some example utterances:

```
navigate/O to/O my/O home/O
navigate/O to/O Oxford/FAC Street/FAC
```

Obviously, we want NER to perform better and recognize location entities, such as street names, district names, and other location names, such as home, work, and office. Now, we'll feed our examples to the NER component and will do more training. We will train NER in three steps:

1. First, we'll disable all the other statistical pipeline components, including the POS tagger and the dependency parser.

2. We'll feed our domain examples to the training procedure.

3. We'll evaluate the new NER model.

Also, we will learn how to do the following:

- Save the updated NER model to disk.

- Read the updated NER model when we want to use it.

Let's get started and dive into training the NER model procedure. As we pointed out in the preceding list, we'll train the NER model in several steps. We'll start with the first step, disabling the other statistical models of the spaCy NLP pipeline.

Disabling the other statistical models

Before starting the training procedure, we disable the other pipeline components, hence we train **only** the intended component. The following code segment disables all the pipeline components except NER. We call this code block before starting the training procedure:

```
other_pipes = [pipe for pipe in nlp.pipe_names if pipe !=
'ner']
  nlp.disable_pipes(*other_pipes)
```

Another way of writing this code is as follows:

```
other_pipes = [pipe for pipe in nlp.pipe_names if pipe !=
'ner']
  with nlp.disable_pipes(*other_pipes):
    # training code goes here
```

In the preceding code block, we made use of the fact that `nlp.disable_pipes` returns a context manager. Using a `with` statement makes sure that our code releases the allocated sources (such as file handlers, database locks, or multiple threads). If you're not familiar with statements, you can read more at this Python tutorial: `https://book.pythontips.com/en/latest/context_managers.html`.

We have completed the first step of the training code. Now, we are ready to make the model training procedure.

Model training procedure

As we mentioned in *Chapter 3*, *Linguistic Features*, in the *Introducing named entity recognition* section, spaCy's NER model is a neural network model. To train a neural network, we need to configure some parameters as well as provide training examples. Each prediction of the neural network is a sum of its **weight** values; hence, the training procedure adjusts the weights of the neural network with our examples. If you want to learn more about how neural networks function, you can read the excellent guide at `http://neuralnetworksanddeeplearning.com/`.

In the training procedure, we'll go over the training set *several times* and show each example several times (one iteration is called one **epoch**) because showing an example only once is not enough. At each iteration, we shuffle the training data so that the order of the training data does not matter. This shuffling of training data helps train the neural network thoroughly.

In each epoch, the training code updates the weights of the neural network with a small number. Optimizers are functions that update the neural network weights subject to a loss. At epoch, a loss value is calculated by comparing the actual label with the neural network's current output. Then, the optimizer function can update the neural network's weight with respect to this loss value.

In the following code, we used the **stochastic gradient descent** (**SGD**) algorithm as the optimizer. SGD itself is also an iterative algorithm. It aims to minimize a function (for neural networks, we want to minimize the loss function). SGD starts from a random point on the loss function and travels down its slope in steps until it reaches the lowest point of that function. If you want to learn more about SGD, you can visit Stanford's excellent neural network class at `http://deeplearning.stanford.edu/tutorial/supervised/OptimizationStochasticGradientDescent/`.

Putting it all altogether, here's the code to train spaCy's NER model for the navigation domain. Let's go step by step:

1. In the first three lines, we make the necessary imports. `random` is a Python library that includes methods for pseudo-random generators for several distributions, including uniform, gamma, and beta distributions. In our code, we'll use `random.shuffle` to shuffle our dataset. `shuffle` shuffles sequences into place:

    ```
    import random
    import spacy
    from spacy.training import Example
    ```

2. Next, we will create a language pipeline object, `nlp`:

    ```
    nlp = spacy.load("en_core_web_md")
    ```

3. Then, we will define our navigation domain training set sentences. Each example contains a sentence and its annotation:

    ```
    trainset = [
                ("navigate home", {"entities": [(9,13,
        "GPE")]}),
                ("navigate to office", {"entities": [(12,18,
        "GPE")]}),
                ("navigate", {"entities": []}),
                ("navigate to Oxford Street", {"entities":
        [(12, 25, "GPE")]})
                ]
    ```

4. We want to iterate our data 20 times, hence the number of epochs is 20:

```
epochs = 20
```

5. In the next 2 lines, we disable the other pipeline components and leave NER for training. We use `with` statement to invoke `nlp.disable_pipe` as a context manager:

```
other_pipes = [pipe for pipe in nlp.pipe_names if pipe !=
'ner']
with nlp.disable_pipes(*other_pipes):
```

6. We create an `optimizer` object, as we discussed previously. We'll feed this `optimizer` object to the training method as a parameter:

```
optimizer = nlp.create_optimizer()
```

7. Then, for each epoch, we will shuffle our dataset by `random.shuffle`:

```
for i in range(epochs):
    random.shuffle(trainset)
```

8. For each example sentence in the dataset, we will create an `Example` object from the sentence and its annotation:

```
example = Example.from_dict(doc, annotation)
```

9. We will feed the `Example` object and `optimizer` object to `nlp.update`. The actual training method is `nlp.update`. This is the place where the NER model gets trained:

```
nlp.update([example], sgd=optimizer)
```

10. Once the epochs are complete, we save the newly trained NER component to disk under a directory called `navi_ner`:

```
ner = nlp.get_pipe("ner")
ner.to_disk("navi_ner")'
```

`nlp.update` outputs a loss value each time it is called. After invoking this code, you should see an output similar to the following screenshot (the loss values might be different):

```
{'ner': 1.9998691082724183}
{'ner': 0.0}
{'ner': 1.7909310894735695}
{'ner': 1.998622537366741}
{'ner': 0.0}
{'ner': 0.9735670930683007}
{'ner': 3.273535776315839e-05}
{'ner': 1.3424393931363738}
{'ner': 0.0}
{'ner': 1.5764574679778889}
{'ner': 2.0439539852645794e-05}
{'ner': 0.04997271008937787}
{'ner': 0.017373304935972556}
{'ner': 5.558857174037257e-05}
{'ner': 4.840094106839388e-06}
{'ner': 0.0}
{'ner': 6.678458655429154e-05}
{'ner': 1.342478230625943e-05}
{'ner': 0.0}
{'ner': 9.123670459132427e-07}
{'ner': 7.241658629211778e-07}
{'ner': 0.0}
{'ner': 2.1249866972522513e-05}
{'ner': 3.0433474562302498e-05}
{'ner': 3.52999521488484e-08}
{'ner': 0.0}
{'ner': 1.75469403973347027e-05}
{'ner': 5.0824145162164935e-06}
{'ner': 4.826914962313689e-07}
{'ner': 0.0}
{'ner': 2.789162502295911e-07}
{'ner': 2.2365957666643788e-07}
{'ner': 1.1323698455023568e-06}
{'ner': 0.0}
{'ner': 1.731425174003251e-06}
```

Figure 7.4 – NER training's output

That's it! We trained the NER component for the navigation domain! Let's try some example sentences and see whether it really worked.

Evaluating the updated NER

Now we can test our brand-new updated NER component. We can try some examples with synonyms and paraphrases to test whether the neural network really learned the navigation domain, instead of memorizing our examples. Let's see how it goes:

1. These are the training sentences:

    ```
    navigate home
    navigate to office
    navigate
    navigate to Oxford Street
    ```

2. Let's use the synonym house for home and also add two more words to to my:

    ```
    doc= nlp("navigate to my house")
    doc.ents
    (house,)
    doc.ents[0].label_
    'GPE'
    ```

3. It worked! House is recognized as a GPE type entity. How about we replace navigate with a similar verb, drive me, and create a paraphrase of the first example sentence:

    ```
    doc= nlp("drive me to home")
    doc.ents
    (home,)
    doc.ents[0].label_
    'GPE'
    ```

4. Now, we try a slightly different sentence. In the next sentence, we won't use a synonym or paraphrase. We'll replace Oxford Street with a district name, Soho. Let's see what happens this time:

    ```
    doc= nlp("navigate to Soho")
    doc.ents
    (Soho,)
    doc.ents[0].label_
    'GPE'
    ```

5. As we remarked before, we updated the statistical model, hence, the NER model didn't forget about the entities it already knew. Let's do a test with another entity type to see whether the NER model really didn't forget the other entity types:

```
doc = nlp("I watched a documentary about Lady Diana.")
doc.ents
(Lady Diana,)
doc.ents[0].label_
'PERSON'
```

Great! spaCy's neural networks can recognize not only synonyms but entities of the same type. This is one of the reasons why we use spaCy for NLP. Statistical models are incredibly powerful.

In the next section, we'll learn how to save the model we trained and load a model into our Python scripts.

Saving and loading custom models

In the preceding code segment, we already saw how to serialize the updated NER component as follows:

```
ner = nlp.get_pipe("ner")
ner.to_disk("navi_ner")
```

We serialize models so that we can upload them in other Python scripts whenever we want. When we want to upload a custom-made spaCy component, we perform the following steps:

```
import spacy

nlp = spacy.load("en_core_web_md", disable=["ner"])
ner = nlp.create_pipe("ner")
ner.from_disk("navi_ner")
nlp.add_pipe(ner, "navi_ner")
print(nlp.meta['pipeline'])
['tagger', 'parser', 'navi_ner']
```

Here are the steps that we follow:

1. We first load the pipeline components without the NER, because we want to add our custom NER. This way, we make sure that the default NER doesn't override our custom NER component.

2. Next, we create an NER pipeline component object. Then we load our custom NER component from the directory we serialized to this newly created component object.

3. We then add our custom NER component to the pipeline.

4. We print the metadata of the pipeline to make sure that loading our custom component worked.

Now, we also learned how to serialize and load custom components. Hence, we can move forward to a bigger mission: training a spaCy statistical model from scratch. We'll again train the NER component, but this time we'll start from scratch.

Training a pipeline component from scratch

In the previous section, we saw how to update the existing NER component according to our data. In this section, we will create a brand-new NER component for the medicine domain.

Let's start with a small dataset to understand the training procedure. Then we'll be experimenting with a real medical NLP dataset. The following sentences belong to the medicine domain and include medical entities such as drug and disease names:

```
Methylphenidate/DRUG is effectively used in treating children
with epilepsy/DISEASE and ADHD/DISEASE.
```
```
Patients were followed up for 6 months.
```
```
Antichlamydial/DRUG antibiotics/DRUG may be useful for curing
coronary-artery/DISEASE disease/DISEASE.
```

The following code block shows how to train an NER component from scratch. As we mentioned before, it's better to create our own NER rather than updating spaCy's default NER model as medical entities are not recognized by spaCy's NER component at all. Let's see the code and also compare it to the code from the previous section. We'll go step by step:

1. In the first three lines, we made the necessary imports. We imported `spacy` and `spacy.training.Example`. We also imported `random` to shuffle our dataset:

```
import random
import spacy
from spacy.training import Example
```

2. We defined our training set of three examples. For each example, we included a sentence and its annotated entities:

```
train_set = [
                ("Methylphenidate is effectively used in
treating children with epilepsy and ADHD.", {"entities":
[(0, 15, "DRUG"), (62, 70, "DISEASE"), (75, 79,
"DISEASE")]}),
                ("Patients were followed up for 6
months.", {"entities": []}),
                ("Antichlamydial antibiotics may be
useful for curing coronary-artery disease.", {"entities":
[(0, 26, "DRUG"), (52, 75, "DIS")]})
]
```

3. We also listed the set of entities we want to recognize – DIS for disease names, and DRUG for drug names:

```
entities = ["DIS", "DRUG"]
```

4. We created a blank model. This is different from what we did in the previous section. In the previous section, we used spaCy's pre-trained English language pipeline:

```
nlp = spacy.blank("en")
```

5. We also created a blank NER component. This is also different from the previous section's code. We used the pre-trained NER component in the previous section:

```
ner = nlp.add_pipe("ner")
ner
```

```
<spacy.pipeline.ner.EntityRecognizer object at
0x7f54b50044c0>
```

6. Next, we add each medical label to the blank NER component by using `ner.add_label`:

```
for ent in entities:
    ner.add_label(ent)
```

7. We define the number of epochs as `25`:

```
epochs = 25
```

8. The next two lines disable the other components other than the NER:

```
other_pipes = [pipe for pipe in nlp.pipe_names if pipe !=
'ner']
with nlp.disable_pipes(*other_pipes):
```

9. We created an optimizer object by calling `nlp.begin_training`. This is different from the previous section. In the previous section, we created an optimizer object by calling `nlp.create_optimizer`, so that NER doesn't forget the labels it already knows. Here, `nlp.begin_training` initializes the NER model's weights with 0, hence, the NER model forgets everything it learned before. This is what we want; we want a blank NER model to train from scratch:

```
optimizer = nlp.begin_training()
```

10. For each epoch, we shuffle our small training set and train the NER component with our examples:

```
for i in range(25):
    random.shuffle(train_set)
    for text, annotation in train_set:
        doc = nlp.make_doc(text)
        example = Example.from_dict(doc, annotation)
        nlp.update([example], sgd=optimizer)
```

Here's what this code segment outputs (the loss values may be different):

```
'ner': 0.0}
'ner': 11.142856240272522}
'ner': 10.957241237163544}
'ner': 10.750467658042908}
'ner': 10.106508314609528}
'ner': 0.0}
'ner': 9.044422030448914}
'ner': 0.0}
'ner': 6.249797374010086}
'ner': 5.534269496798515}
'ner': 3.387359671294689}
'ner': 0.0}
'ner': 0.0}
'ner': 1.8647982060501818}
'ner': 3.7477317565935664}
'ner': 18.449425548315048}
'ner': 0.0}
'ner': 2.5493467298983887}
'ner': 0.0}
'ner': 2.0483578685143584}
'ner': 10.296224442208768}
'ner': 1.541039283369173}
'ner': 5.376370556281472}
'ner': 0.0}
'ner': 6.661640930324211}
```

Figure 7.5 – Loss values during training

Did it really work? Let's test the newly trained NER component:

```
doc = nlp("I had a coronary disease.")
doc.ents
(coronary disease,)
doc.ents[0].label_
'DIS'
```

Great – it worked! Let's also test some negative examples, entities that are recognized by spaCy's pre-trained NER model but not ours:

```
doc = nlp("I met you at Trump Tower.")
doc.ents
()
doc = nlp("I meet you at SF.")
doc.ents
()
```

This looks good, too. Our brand new NER recognizes only medical entities. Let's visualize our first example sentence and see how displaCy exhibits new entities:

```
from spacy import displacy
doc = nlp("I had a coronary disease.")
displacy.serve(doc, style="ent")
```

This code block generates the following visualization:

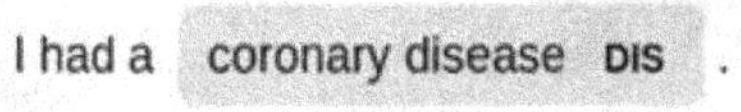

I had a   coronary disease DIS .

Figure 7.6 – Visualization of the example sentence

We successfully trained the NER model on small datasets. Now it's time to work with a real-world dataset. In the next section, we'll dive into processing a very interesting dataset regarding a hot topic; mining Corona medical texts.

Working with a real-world dataset

In this section, we will train on a real-world corpus. We will train an NER model on the CORD-19 corpus provided by the *Allen Institute for AI* (https://allenai.org/). This is an open challenge for text miners to extract information from this dataset to help medical professionals around the world fight against Corona disease. CORD-19 is an open source dataset that is collected from over 500,000 scholarly articles about Corona disease. The training set consists of 20 annotated medical text samples:

1. Let's get started by having a look at an example training text:

    ```
    The antiviral drugs amantadine and rimantadine inhibit
    a viral ion channel (M2 protein), thus inhibiting
    replication of the influenza A virus.[86] These drugs are
    sometimes effective against influenza A if given early
    ```

```
in the infection but are ineffective against influenza
B viruses, which lack the M2 drug target. [160] Measured
resistance to amantadine and rimantadine in American
isolates of H3N2 has increased to 91% in 2005. [161]
This high level of resistance may be due to the easy
availability of amantadines as part of over-the-counter
cold remedies in countries such as China and Russia, [162]
and their use to prevent outbreaks of influenza in farmed
poultry. [163] [164] The CDC recommended against using M2
inhibitors during the 2005-06 influenza season due to
high levels of drug resistance. [165]
```

As we see from this example, real-world medical text can be quite long, and it can include many medical terms and entities. Nouns, verbs, and entities are all related to the medicine domain. Entities can be numbers (91%), number and units (100 ng/ ml, 25 microg/ml), number-letter combinations (H3N2), abbreviations (CDC), and also compound words (qRT-PCR, PE-labeled).

The medical entities come in several shapes (numbers, number and letter combinations, and compounds) as well as being very domain-specific. Hence, a medical text is very different from everyday spoken/written language and definitely needs custom training.

2. Entity labels can be compound words as well. Here's the list of entity types that this corpus includes:

```
Pathogen
MedicalCondition
Medicine
```

We transformed the dataset so that it's ready to use with spaCy training. The dataset is available under the book's GitHub repository: https://github.com/ PacktPublishing/Mastering-spaCy/tree/main/Chapter07/data.

3. Let's go ahead and download the dataset. Type the following command into your terminal:

```
$wget
https://github.com/PacktPublishing/Mastering-spaCy/blob/
main/Chapter07/data/corona.json
```

This will download the dataset into your machine. If you wish, you can manually download the dataset from GitHub, too.

4. Now, we'll preprocess the dataset a bit to recover some format changes that happened while dumping the dataset as json:

```
import json
with open("data/corona.json") as f:
    data = json.loads(f.read())


TRAIN_DATA = []
for (text, annotation) in data:
    new_anno = []
    for anno in annotation["entities"]:
        st, end, label = anno
        new_anno.append((st, end, label))
    TRAIN_DATA.append((text, {"entities": new_anno}))
```

This code segment will read the dataset's JSON file and format it according to the spaCy training data conventions.

5. Next, we'll do the statistical model training:

a) First, we'll do the related imports:

```
import random
import spacy
from spacy.training import Example
```

b) Secondly, we'll initialize a blank spaCy English model and add an NER component to this blank model:

```
nlp = spacy.blank("en")
ner = nlp.add_pipe("ner")
print(ner)
print(nlp.meta)
```

c) Next, we define the labels we'd like the NER component to recognize and introduce these labels to it:

```
labels = ['Pathogen', 'MedicalCondition', 'Medicine']
for ent in labels:
    ner.add_label(ent)
print(ner.labels)
```

d) Finally, we're ready to define the training loop:

```
epochs = 100
other_pipes = [pipe for pipe in nlp.pipe_names if pipe !=
'ner']
with nlp.disable_pipes(*other_pipes):
  optimizer = nlp.begin_training()
  for i in range(100):
    random.shuffle(TRAIN_DATA)
    for text, annotation in TRAIN_DATA:
        doc = nlp.make_doc(text)
        example = Example.from_dict(doc, annotation)
        nlp.update([example], sgd=optimizer)
```

This code block is identical to the training code from the previous section, except for the value of the epochs variable. This time, we iterated for 100 epochs, because the entity types, entity values, and the training sample text are semantically more complicated. We recommend you do at least 500 iterations for this dataset if you have the time. 100 iterations over the data are sufficient to get good results, but 500 iterations will take the performance further.

6. Let's visualize some sample texts to see how our newly trained medical NER model handled the medical entities. We'll visualize our medical entities with displaCy code:

```
from spacy import displacy
doc = nlp("One of the bacterial diseases with the highest
disease burden is tuberculosis, caused by Mycobacterium
tuberculosis bacteria, which kills about 2 million people
a year.")
displacy.serve(doc, style="ent")
```

The following screenshot highlights two entities – tuberculosis and the name of the bacteria that causes it as the pathogen entity:

One of the bacterial diseases with the highest disease burden is tuberculosis **MedicalCondition** , caused by Mycobacterium tuberculosis **Pathogen** bacteria, which kills about 2 million people a year.

Figure 7.7 – Highlighted entities of the sample medical text

7. This time, let's look at entities of a text concerning pathogenic bacteria. This sample text contains many entities, including several diseases and pathogen names. All the disease names, such as pneumonia, tetanus, and leprosy, are correctly extracted by our medical NER model. The following displaCy code highlights the entities:

```
doc2 = nlp("Pathogenic bacteria contribute to other
globally important diseases, such as pneumonia, which
can be caused by bacteria such as Streptococcus and
Pseudomonas, and foodborne illnesses, which can be
caused by bacteria such as Shigella, Campylobacter, and
Salmonella. Pathogenic bacteria also cause infections
such as tetanus, typhoid fever, diphtheria, syphilis, and
leprosy. Pathogenic bacteria are also the cause of high
infant mortality rates in developing countries.")
displacy.serve(doc2, style="ent")
```

Here is the visual generated by the preceding code block:

Pathogenic bacteria **Pathogen** contribute to other globally important diseases, such as pneumonia **MedicalCondition** , which can be caused by bacteria such as Streptococcus **Pathogen** and Pseudomonas **Pathogen** , and foodborne illnesses, which can be caused by bacteria such as Shigella **Pathogen** , Campylobacter **Pathogen** , and Salmonella **Pathogen** . Pathogenic bacteria also cause infections such as tetanus **MedicalCondition** , typhoid fever **MedicalCondition** , diphtheria **MedicalCondition** , syphilis **MedicalCondition** , and leprosy **MedicalCondition** . Pathogenic bacteria are also the cause of high infant mortality rates in developing countries.

Figure 7.8 – Sample text with disease and pathogen entities highlighted

Looks good! We successfully trained spaCy's NER model for the medicine domain and now the NER can extract information from medical text. This concludes our section. We learned how to train a statistical pipeline component as well as prepare the training data and test the results. These are great steps in both mastering spaCy and machine learning algorithm design.

Summary

In this chapter, we explored how to customize spaCy statistical models according to our own domain and data. First, we learned the key points of deciding whether we really need custom model training. Then, we went through an essential part of statistical algorithm design – data collection, and labeling.

Here we also learned about two annotation tools – Prodigy and Brat. Next, we started model training by updating spaCy's NER component with our navigation domain data samples. We learned the necessary model training steps, including disabling the other pipeline components, creating example objects to hold our examples, and feeding our examples to the training code.

Finally, we learned how to train an NER model from scratch on a small toy dataset and on a real medical domain dataset.

With this chapter, we took a step into the statistical NLP playground. In the next chapter, we will take more steps in statistical modeling and learn about text classification with spaCy. Let's move forward and see what spaCy brings us!

8
Text Classification with spaCy

This chapter is devoted to a very basic and popular task of NLP: text classification. You will first learn how to train spaCy's text classifier component, `TextCategorizer`. For this, you will learn how to prepare data and feed the data to the classifier; then we'll proceed to train the classifier. You'll also practice your new `TextCategorizer` skills on a popular dataset for sentiment analysis.

Next, you will also do text classification with the popular framework TensorFlow's Keras API together with spaCy. You will learn the basics of neural networks, sequential data modeling with LSTMs, and how to prepare text for machine learning tasks with Keras's text preprocessing module. You will also learn how to design a neural network with `tf.keras`.

Following that, we will then make an end-to-end text classification experiment, from data preparation to preprocessing text with Keras `Tokenizer`, neural network designing, model training, and interpreting the classification results. That's a whole package of machine learning!

In this chapter, we're going to cover the following main topics:

- Understanding the basics of text classification
- Training the spaCy text classifier

- Sentiment analysis with spaCy
- Text classification with spaCy and Keras

Technical requirements

The code in the sections *Training the spaCy text classifier* and *Sentiment analysis with spaCy* is spaCy v3.0 compatible.

The section *Text classification with spaCy and Keras* requires the following Python libraries:

- TensorFlow >=2.2.0
- NumPy
- pandas
- Matplotlib

You can install the latest version of these libraries with `pip` as follows:

```
pip install tensorflow
pip install numpy
pip install pandas
pip install matplotlib
```

We also use Jupyter notebooks in the last two sections. You can follow the instructions on the Jupyter website (`https://jupyter.org/install`) to install the Jupyter notebook onto your system. If you don't want to use notebooks, you can copy-paste code as Python code as well.

You can find the chapter code and data files in the book's GitHub repository at `https://github.com/PacktPublishing/Mastering-spaCy/tree/main/Chapter08`.

Let's get started with spaCy's text classifier component first, then we'll transition to designing our own neural network.

Understanding the basics of text classification

Text classification is the task of assigning a set of predefined labels to text. Given a set of predefined classes and some text, you want to understand which predefined class this text falls into. We have to determine the classes ourselves by the nature of our data before starting the classification task. For example, a customer review can be positive, negative, or neutral.

Text classifiers are used for detecting spam emails in your mailbox, determining the sentiment of customer's reviews, understanding customer's intent, sorting customer's complaint tickets, and so on.

Text classification is a fundamental task of NLP. It is gaining importance in the business world, as it enables businesses to automate their processes. One immediate example is spam filters. Every day, users receive many spam emails but most of the time never see these emails and don't get any notifications because spam filters save the users from bothering about irrelevant emails and from spending time deleting these emails.

Text classifiers can come in different flavors. Some classifiers focus on the overall emotion of the text, some classifiers focus on detecting the language of the text, and some classifiers focus on only some words of the text, such as verbs. The following are some of the most common types of text classification and their use cases:

- **Topic detection**: Topic detection is the task of understanding the topic of a given text. For example, the text in a customer email could be asking about a refund, asking for a past bill, or simply complaining about the customer service.

- **Sentiment analysis**: Sentiment analysis is the task of understanding whether the text contains positive or negative emotions about a given subject. Sentiment analysis is used often to analyze customer reviews about products and services.

- **Language detection**: Language detection is the first step of many NLP systems, such as machine translation.

The following figure shows a text classifier for a customer service automation system:

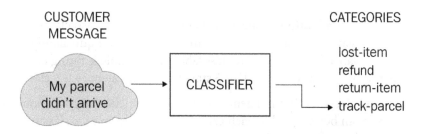

Figure 8.1 – Topic detection is used to label a customer complaint with a predefined label

Coming to the technical details, text classification is a *supervised* machine learning task. It means that the classifier can predict the class label of a text based on *example* input text-class label pairs. Hence, to train a text classifier, we need a *labeled dataset*. A labeled dataset is basically a list of text-label pairs. Here is an example dataset of five training sentences with their labels:

```
This shampoo is great for hair.
                  POSITIVE
I loved this shampoo, best product ever!
        POSITIVE
My hair has never been better, great product. POSITIVE
This product make my scalp itchy.
                NEGATIVE
Not the best quality for this price.
                NEGATIVE
```

Then we train the classifier by showing the text and the corresponding class labels to the classifier. When the classifier sees new text that was not in the training text, it then predicts the class label of this unseen text based on the examples it saw during the training phase. The output of a text classifier is *always* a class label.

Text classification can also be divided into three categories depending on the number of classes used:

- **Binary text classification** means that we want to categorize our text into two classes.

- **Multiclass text classification** means that there are more than two classes. Each class is mutually exclusive – one text can belong to one class only. Equivalently, a training instance can be labeled with only one class label. An example is rating customer reviews. A review can have 1, 2, 3, 4, or 5 stars (each star category is a class).

- **Multilabel text classification** is a generalization of multiclass classification, where multiple labels can be assigned to each example text. For example, classifying toxic social media messages is done with multiple labels. This way, our model can distinguish different levels of toxicity. Class labels are typically toxic, severe toxic, insult, threat, obscenity. A message can include both insults and threats, or be classed as insult, toxicity, and obscenity, and so on. Hence for this problem, using multiple classes is more suitable.

Labels are the name of the classes we want to see as the output. A class label can be categorical (string) or numerical (a number). Here are some commonly used class labels:

- For sentiment analysis, we usually use the class labels positive and negative. Their abbreviations, pos and neg, are also commonly used. Binary class labels are popular as well – 0 means negative sentiment and 1 means positive sentiment.

- The same applies to binary classification problems. We usually use 0-1 for class labels.

- For multiclass and multilabel problems, we usually name the classes with a meaningful name. For a movie genre classifier, we can use the labels family, international, Sunday evening, Disney, action, and so on. Numbers are used as labels as well. For a five-class classification problem, we can use the labels 1, 2, 3, 4, and 5.

Now we've covered the basic concepts of text classification, let's do some coding! In the next section, we'll explore how to train spaCy's text classifier component.

Training the spaCy text classifier

In this section, we will learn about the details of spaCy's text classifier component TextCategorizer. In *Chapter 2, Core Operations with spaCy*, we saw that the spaCy NLP pipeline consists of components. In *Chapter 3, Linguistic Features*, we learned about the essential components of the spaCy NLP pipeline, which are the sentence tokenizer, POS tagger, dependency parser, and **named entity recogition** (**NER**).

TextCategorizer is an optional and trainable pipeline component. In order to train it, we need to provide examples and their class labels. We first add TextCategorizer to the NLP pipeline and then do the training procedure. *Figure 8.2* shows where exactly the TextCategorizer component lies in the NLP pipeline; this component comes after the essential components. In the following diagram, **textcat** refers to the TextCategorizer component.

Figure 8.2 – TextCategorizer in the nlp pipeline

A neural network architecture lies behind spaCy's TextCategorizer. TextCategorizer provides us with user-friendly and end-to-end approaches to train the classifier, so we don't have to deal directly with the neural network architecture. We'll design our own neural network architecture in the upcoming *Text classification with spaCy and Keras* section. After looking at the architecture, we're ready to dive into TextCategorizer code. Let's get to know TextCategorizer class first.

Getting to know TextCategorizer class

Now let's get to know the TextCategorizer class in detail. First of all, we import TextCategorizer from the pipeline components:

```
from spacy.pipeline.textcat import DEFAULT_SINGLE_TEXTCAT_MODEL
```

TextCategorizer is available in two flavors, single-label classifier and multilabel classifier. As we remarked in the previous section, a multilabel classifier can predict more than one class. A single-label classifier predicts only one class for each example and classes are mutually exclusive. The preceding import line imports the single-label classifier and the following code imports the multilabel classifier:

```
from spacy.pipeline.textcat_multilabel import DEFAULT_MULTI_
TEXTCAT_MODEL
```

Next, we need to provide a configuration to the TextCategorizer component. We provide two parameters here, a threshold value and a model name (either Single or Multi depending on the classification task). TextCategorizer internally generates a probability for each class and a class is assigned to the text if the probability of this class is higher than the threshold value.

A traditional threshold value for text classification is 0.5, however, if you want to make a prediction with higher confidence, you can make the threshold higher, such as 0.6, 0.7, or 0.8.

Putting it altogether, we can add a single-label TextCategorizer component to the nlp pipeline as follows:

```
from spacy.pipeline.textcat import DEFAULT_SINGLE_TEXTCAT_MODEL
config = {
    "threshold": 0.5,
    "model": DEFAULT_SINGLE_TEXTCAT_MODEL
}
textcat = nlp.add_pipe("textcat", config=config)
```

```
textcat
<spacy.pipeline.textcat.TextCategorizer object at
0x7f0adf004e08>
```

Adding a multilabel component to the `nlp` pipeline is similar:

```
from spacy.pipeline.textcat_multilabel import
DEFAULT_MULTI_TEXTCAT_MODEL
config = {
    "threshold": 0.5,
    "model": DEFAULT_MULTI_TEXTCAT_MODEL
}
textcat = nlp.add_pipe("textcat_multilabel", config=config)
textcat
<spacy.pipeline.textcat.TextCategorizer object at
0x7f0adf004e08>
```

In the last line of each of the preceding code blocks, we added a `TextCategorizer` pipeline component to the nlp pipeline object. The newly created `TextCategorizer` component is captured by the `textcat` variable. We're ready to train the `TextCategorizer` component now. The training code looks quite similar to the NER component training code from *Chapter 7, Customizing spaCy Models*, except for some minor details.

Formatting training data for the TextCategorizer

Let's start our code by preparing a small training set. We'll prepare a customer sentiment dataset for binary text classification. The label will be called `sentiment` and can obtain two possible values, 0 and 1 corresponding to negative and positive sentiment. The following training set contains 6 examples, 3 being positive and 3 being negative:

```
train_data = [
    ("I loved this product, very easy to use.", {"cats":
{"sentiment": 1}}),
    ("I'll definitely purchase again. I recommend this
product.", {"cats": {"sentiment": 1}}),
    ("This is the best product ever. I loved the scent and the
feel. Will buy again.", {"cats": {"sentiment": 1}}),
    ("Disappointed. This product didn't work for me at all",
{"cats": {"sentiment": 0}}),
```

```
    ("I hated the scent. Won't buy again", {"cats":
{"sentiment": 0}}),
    ("Truly horrible product. Very few amount of product for a
high price. Don't recommend.", {"cats": {"sentiment": 0}})
]
```

Each training example is a tuple of a text and a nested dictionary. The dictionary contains the class label in a format that spaCy recognizes. The cts field means the categories. Then we include the class label sentiment and its value. The value should always be a floating-point number.

In the code, we will introduce the class label we choose to the TextCategorizer component. Let's see the complete code. First, we do the necessary imports:

```
import random
import spacy
from spacy.training import Example
from spacy.pipeline.textcat import DEFAULT_SINGLE_TEXTCAT_MODEL
```

We imported the built-in library random for shuffling our dataset. We imported spacy as usual, and we imported Example to prepare the training examples in spaCy format. In the last line of the code block, we imported a text categorizer model.

Next, we'll do the pipeline and TextCategorizer component initialization:

```
nlp = spacy.load("en_core_web_md")
config = {
    "threshold": 0.5,
    "model": DEFAULT_SINGLE_TEXTCAT_MODEL
}
textcat = nlp.add_pipe("textcat", config=config)
```

Now, we'll do some work on the newly created TextCategorizer component, textcat. We'll introduce our label sentiment to the TextCategorizer componenet by calling add_label. Then, we need to initialize this component with our examples. This step is different than what we did in the NER training code in *Chapter 7, Customizing spaCy Models*.

The reason is NER is an essential component, hence it's initialized by the pipeline always. TextCategorizer is an optional component, and it comes as a blank statistical model. The following code adds our label to the `TextCategorizer` component and then initializes the `TextCategorizer` model's weights with the training examples:

```
textcat.add_label("sentiment")
train_examples = [Example.from_dict(nlp.make_doc(text), label)
for text,label in train_data]
textcat.initialize(lambda: train_examples, nlp=nlp)
```

Note that we feed the examples to `textcat.initialize` as Example objects. Recall from *Chapter 7, Customizing spaCy Models*, that spaCy training methods always work with Example objects.

Defining the training loop

We're ready to define the training loop. First of all, we'll disable other pipe components so that only `textcat` will be trained. Second, we will create an optimizer object by calling `resume_training`, keeping the weights of the existing statistical models. For each epoch, we go over training examples one by one and update the weights of `textcat`. We go over the data for 20 epochs. The following code defines the training loop:

```
epochs=20
with nlp.select_pipes(enable="textcat"):
  optimizer = nlp.resume_training()
  for i in range(epochs):
    random.shuffle(train_data)
    for text, label in train_data:
      doc = nlp.make_doc(text)
      example = Example.from_dict(doc, label)
      nlp.update([example], sgd=optimizer)
```

That's it! With this relatively short code segment, we trained a text classifier! Here's the output on my machine (your loss values might be different):

```
{'textcat': 0.25}
{'textcat': 0.25403276085853577}
{'textcat': 0.23483306169509888}
{'textcat': 0.2338048851770401}
{'textcat': 0.21322503685951233}
{'textcat': 0.19576627016067505}
{'textcat': 0.17215843498706818}
{'textcat': 0.17311060428619385}
{'textcat': 0.128489151597023}
{'textcat': 0.13611891865730286}
{'textcat': 0.11497252434492111}
{'textcat': 0.07584841549396515}
{'textcat': 0.05945904180407524}
{'textcat': 0.07257045060396194}
{'textcat': 0.03416129946708679}
{'textcat': 0.04692266136407852}
{'textcat': 0.019232971593737602}
{'textcat': 0.029277725145220757}
{'textcat': 0.010624242015182972}
```

Figure 8.3 – Loss values at each epoch

Testing the new component

Let's test the new text categorizer component. The `doc.cats` property holds the class labels:

```
doc2 = nlp("This product sucks")
doc2.cats
{'sentiment': 0.09907063841819763}
doc3 = nlp("This product is great")
doc3.cats
{'sentiment': 0.9740120000120339}
```

Great! Our small dataset successfully trained the spaCy text classifier for a binary text classification problem, indeed a sentiment analysis task. Now, we'll see how to do multilabel classification with spaCy's `TextCategorizer`.

Training TextCategorizer for multilabel classification

Recall from the first section that multilabel classification means the classifier can predict more than one label for an example text. Naturally, the classes are not mutually exclusive at all. In order to train a multilabel classifier, we need to provide a dataset that contains examples that have more than one label.

To train spaCy's `TextCategorizer` for multilabel classification, we'll again start by building a small training set. This time, we'll form a set of movie reviews, where the labels are FAMILY, THRILLER, and SUNDAY_EVENING. Here is our small dataset:

```
train_data = [
    ("It's the perfect movie for a Sunday evening.", {"cats":
{"SUNDAY_EVENING": True}}),
    ("Very good thriller", {"cats": {"THRILLER": True}}),
    ("A great movie for the kids and all the family"   ,
{"cats": {"FAMILY": True}}),
    ("An ideal movie for Sunday night with all the family. My
kids loved the movie.", {"cats": {"FAMILY": True, "SUNDAY_
EVENING":True}}),
    ("A perfect thriller for all the family. No violence, no
drugs, pure action.", {"cats": {"FAMILY": True, "THRILLER":
True}})
]
```

We provided some examples with one label, such as the first example (the first sentence of `train_data`, the second line of the preceding code block), and we also provided examples with more than one label, such as the fourth example of the `train_data`.

We'll make the imports after we've formed the training set:

```
import random
import spacy
from spacy.training import Example
from spacy.pipeline.textcat_multilabel import
DEFAULT_MULTI_TEXTCAT_MODEL
```

Here, the last line is different than the code of the previous section. We imported the multilabel model instead of the single-label model.

Next, we add the multilabel classifier component to the nlp pipeline. Again, pay attention to the pipeline component name – this time, it is `textcat_multilabel`, compared to the previous section's `textcat`:

```
config = {
    "threshold": 0.5,
    "model": DEFAULT_MULTI_TEXTCAT_MODEL
}
textcat = nlp.add_pipe("textcat_multilabel", config=config)
```

Adding the labels to the `TextCategorizer` component and initializing the model is similar to the *Training the spaCy text classifier* section. This time, we'll add three labels instead of one:

```
labels = ["FAMILY", "THRILLER", "SUNDAY_EVENING"]
for label in labels:
    textcat.add_label(label)
train_examples = [Example.from_dict(nlp.make_doc(text), label)
for text,label in train_data]
textcat.initialize(lambda: train_examples, nlp=nlp)
```

We're ready to define the training loop. The code functions are similar to the previous section's code. The only difference is the component name in the first line. Now it's `textcat_multilabel`:

```
epochs=20
with nlp.select_pipes(enable="textcat_multilabel"):
    optimizer = nlp.resume_training()
    for i in range(epochs):
        random.shuffle(train_data)
        for text, label in train_data:
            doc = nlp.make_doc(text)
            example = Example.from_dict(doc, label)
            nlp.update([example], sgd=optimizer)
```

The output should look similar to the output of the previous section, a loss value per epoch. Now, let's test our brand new multilabel classifier:

```
doc2 = nlp("Definitely in my Sunday movie night list")
doc2.cats
{'FAMILY': 0.9044250249862671, 'THRILLER': 0.34271398186683655,
'SUNDAY_EVENING': 0.9801468253135681}
```

Notice that each label admitted a positive probability at the output. Also, the probabilities do not sum up to 1, because they're not mutually exclusive. For this example, the SUNDAY_EVENING and THRILLER label probabilities are predicted correctly, but the FAMILY label probability does not look ideal. This is mainly due to the fact that we didn't provide enough examples. Usually, for multilabel classification problems, the classifier needs more examples than binary classification since the classifier needs to learn more labels.

We've learned how to train spaCy's `TextCategorizer` component for binary text classification and multilabel text classification. Now, we'll train `TextCategorizer` on a real-world dataset for a sentiment analysis problem.

Sentiment analysis with spaCy

In this section, we'll work on a real-world dataset and train spaCy's `TextCategorizer` on this dataset. We'll be working on the Amazon Fine Food Reviews dataset (`https://www.kaggle.com/snap/amazon-fine-food-reviews`) from Kaggle in this chapter. The original dataset is huge, with 100,000 rows. We sampled 4,000 rows. This dataset contains customer reviews about fine food sold on Amazon. Reviews include user and product information, user rating, and text.

You can download the dataset from the book's GitHub repository. Type the following command into your terminal:

```
wget  https://github.com/PacktPublishing/Mastering-spaCy/blob/
main/Chapter08/data/Reviews.zip
```

Alternatively, you can click on the URL in the preceding command and the download will start. You can unzip the zip file with the following:

```
unzip Reviews.zip
```

Alternatively, you can right-click on the ZIP file and choose **Extract here** to inflate the ZIP file.

Exploring the dataset

Now, we're ready to explore the dataset. In this section, we'll be using a Jupyter notebook. If you already have Jupyter installed, you can execute the notebook cells directly. If you don't have the Jupyter Notebook on your system, you can follow the instructions on the Jupyter website (`https://jupyter.org/install`).

Let's do our dataset exploration step by step:

1. First, we'll do the imports for reading and visualizing the dataset:

    ```
    import pandas as pd
    import matplotlib.pyplot as plt
    %matplotlib inline
    ```

2. We'll read the CSV file into a pandas DataFrame and output the shape of the DataFrame:

```
reviews_df=pd.read_csv('data/Reviews.csv')
reviews_df.shape
(3999, 10)
```

3. Next, we examine the rows and the columns of the dataset by printing the first 10 rows:

```
reviews_df.head()
```

The resulting view tells us there are 10 rows including the review text and the review score:

Out[8]:

	Id	ProductId	UserId	ProfileName	HelpfulnessNumerator	HelpfulnessDenominator	Score	Time	Summary	Text
0	50057	B000ER5DFQ	A1ESDLEDR9Y0JX	A. Spencer	1	2	1	1310256000	the garbanzo beans in it give horrible gas	To be fair only one of my twins got gas from t...
1	366917	B001AIQP8M	A324KM3YY1DWQG	danitrice	0	0	5	1251072000	Yummy Lil' Treasures!!	Just recieved our first order of these (they d...
2	214380	B001E5E1XW	A3QCWO53N69HW3	M. A. Vaughan "-_- GOBNOGO-_-"	2	2	5	1276473600	Great Chai	This is a fantastic Chai Masala. I am very pic...
3	178476	B000TIZP5I	AYZ5NG9705AG1	Consumer	0	0	5	1341360000	Celtic Salt worth extra price	Flavorful and has added nutrition! You use le...
4	542504	B000E18CVE	A2LMWCJUF5HZ4Z	Miki Lam "mikilam"	8	11	3	1222732800	mixed feelings	I thought this soup tasted good. I liked the t...

Figure 8.4 – First 10 rows of the reviews dataframe

4. We'll be using the Text and Score columns; hence, we'll drop the other columns that we won't use. We'll also call the dropna() method to drop the rows with missing values::

```
reviews_df = reviews_df[['Text','Score']].dropna()
```

5. We can have a quick look at how the review scores are distributed:

```
ax=reviews_df.Score.value_counts().plot(kind='bar',
colormap='Paired')
plt.show()
```

6. This piece of code calls the `plot` method of `dataframe reviews_df` and exhibits a bar plot:

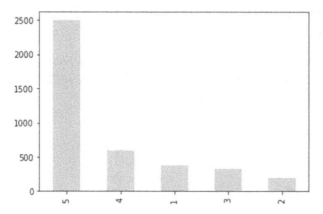

Figure 8.5 – Distribution of review scores

The number of 5-star ratings is quite high; it looks like customers are happy with the food they purchased. However, it might create an imbalance in the training data if a class has significantly more weight than the others.

Class imbalance creates trouble for classification algorithms in general. For example, it is considered as imbalance when a class has significantly more training examples than the other classes (usually a ratio of 1:5 between examples). There are different ways to handle imbalance, one way is **up-sampling/down-sampling**. In down-sampling, we randomly remove training examples from the majority class. In up-sampling, we randomly replicate training example from the minority class. Both methods aim to balance the number of training examples of majority and minority classes.

Here we'll apply another method. We'll combine 1,2,3 star reviews and 4,5 star reviews to get a more balanced dataset.

7. In order to prevent this, we'll treat 1-, 2-, and 3-star ratings as negative and ratings that have more than 4 stars as positive. The following code segment assigns a negative label to all the reviews that have a rating of fewer than 4 stars and a positive label to all the reviews that have a higher rating than 4 stars:

```
reviews_df.Score[reviews_df.Score<=3]=0
reviews_df.Score[reviews_df.Score>=4]=1
```

8. Let's plot the distribution of the ratings again:

```
ax=reviews_df.Score.value_counts().plot(kind='bar',
colormap='Paired')
plt.show()
```

The resulting rating distribution looks much better than *Figure 8.5*. Still, the number of positive reviews is greater, but the number of negative reviews is significant as well, as can be seen from the following graph:

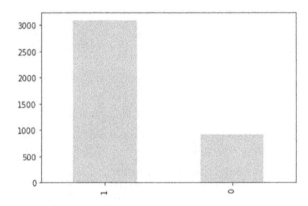

Figure 8.6 – Distribution of positive and negative ratings

After processing the dataset, we reduced it to a two-column dataset with negative and positive ratings. We call `reviews_df.head()` once again and the following is the result we get:

Out[17]:

	Text	Score
0	To be fair only one of my twins got gas from t...	0
1	Just recieved our first order of these (they d...	1
2	This is a fantastic Chai Masala. I am very pic...	1
3	Flavorful and has added nutrition! You use le...	1
4	I thought this soup tasted good. I liked the t...	0

Figure 8.7 – The DataFrame's first four rows

We'll finish our dataset exploration here. We saw the distribution of the review scores and the class labels. The dataset is now ready to be processed. We dropped the unused columns and converted review scores to binary class labels. Let's go ahead and start the training procedure!

Training the TextClassifier component

Now, we're ready to start the training procedure. We'll train a binary text classifier with the multilabel classifier this time. Again, let's go step by step:

1. We start by importing the spaCy classes as follows:

```
import spacy
import random
from spacy.training import Example
from spacy.pipeline.textcat_multilabel import DEFAULT_
MULTI_TEXTCAT_MODEL
```

2. Next, we'll create a pipeline object, nlp, define the classifier configuration, and add the TextCategorizer component to nlp with the following configuration:

```
nlp = spacy.load("en_core_web_md")
 config = {
    "threshold": 0.5,
    "model": DEFAULT_MULTI_TEXTCAT_MODEL
 }
textcat = nlp.add_pipe("textcat_multilabel",
config=config)
```

3. After creating the text classifier component, we'll convert training sentences and ratings into a spaCy usable format. We'll iterate every row of the DataFrame with iterrows(), and for each row we'll extract the Text and Score fields. Then, we'll create a spaCy Doc object from the review text and make a dictionary of the class labels as well. Finally, we will create an Example object and append it to the list of training examples:

```
train_examples = []
for index, row in reviews_df.iterrows():
    text = row["Text"]
    rating = row["Score"]
    label = {"POS": True, "NEG": False} if rating == 1
else {"NEG": True, "POS": False}
    train_examples.append(Example.from_dict(nlp.make_
doc(text), {"cats": label}))
```

4. We'll use POS and NEG labels for positive and negative sentiment, respectively. We'll introduce these labels to the new component and also initialize the component with examples:

```
textcat.add_label("POS")
textcat.add_label("NEG")
textcat.initialize(lambda: train_examples, nlp=nlp)
```

5. We're ready to define the training loop! We went over the training set for two epochs, but you can go over more if you like. The following code snippet will train the new text categorizer component:

```
epochs = 2
with nlp.select_pipes(enable="textcat_multilabel"):
    optimizer = nlp.resume_training()
    for i in range(epochs):
        random.shuffle(train_examples)
        for example in train_examples:
            nlp.update([example], sgd=optimizer)
```

6. Finally, we'll test how the text classifier component works for two example sentences:

```
doc2 = nlp("This is the best food I ever ate")
doc2.cats
{'POS': 0.9553419947624207, 'NEG': 0.061326123774051666}
doc3 = nlp("This food is so bad")
doc3.cats
{'POS': 0.21204468607902527, 'NEG': 0.8010350465774536}
```

Both the NEG and POS labels appear in the prediction result because we used the multilabel classifier. The results look good. The first sentence outputs a very high positive probability, and the second sentence is predicted as negative with a high probability.

We've completed training spaCy's text classifier component. In the next section, we'll dive into the world of a very popular deep learning library, Keras. We'll explore how to write Keras code to do text classification by using another popular machine learning library – TensorFlow's Keras API. Let's go ahead and explore Keras and TensorFlow!

Text classification with spaCy and Keras

In this section, we will learn about methods for blending spaCy with neural networks using another very popular Python deep learning library, **TensorFlow**, and its high-level API, **Keras**.

Deep learning is a broad family of machine learning algorithms that are based on neural networks. **Neural networks** are human brain-inspired algorithms that contain connected layers, which are made from neurons. Each neuron is a mathematical operation that takes its input, multiplies it by its weights, and then passes the sum through the activation function to the other neurons. The following diagram shows a neural network architecture with three layers -- the **input layer**, **hidden layer**, and **output layer**:

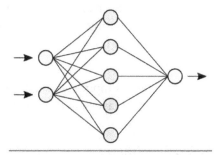

Figure 8.8 – A neural network architecture with three layers

TensorFlow is an end-to-end open source platform for machine learning. TensorFlow might be the most popular deep learning library among research engineers and scientists. It has huge community support and great documentation, available at https://www.tensorflow.org/.

Keras is a high-level deep learning API that can run on top of popular machine learning libraries such as TensorFlow, Theano, and CNTK. Keras is very popular in the research and development world because it supports rapid prototyping and provides a user-friendly API to neural network architectures.

TensorFlow 2 introduced great changes in machine learning methods by tightly integrating with Keras and providing a high-level API, **tf.keras**. TensorFlow 1 was a bit ugly with symbolic graph computations and other low-level computations. With TensorFlow 2, developers can take advantage of Keras' user-friendliness as well as TensorFlow's low-level methods.

Neural networks are commonly used for computer vision and NLP tasks, including object detection, image classification, and scene understanding as well as text classification, POS tagging, text summarization, and natural language generation.

In the following sections, we'll go through the details of a neural network architecture for text classification implemented with `tf.keras`. Throughout this section, we'll use TensorFlow 2 as we stated in the *Technical requirements* section. Let's warm up to neural networks with some neural network basics, and then start building our Keras code.

What is a layer?

A neural network is formed by connecting layers. **Layers** are basically the building blocks of the neural network. A layer consists of several **neurons**, as in *Figure 8.8*.

In *Figure 8.8*, the first layer of this neural network has two layers, and the second layer has six neurons. Each neuron in each layer is connected to all neurons in the next layer. Each layer might have different functionalities; some layers can lower the dimensions of their input, some layers can flatten their input (flattening means collapsing a multidimensional vector into one dimension), and so on. At each layer, we transform the input vectors and feed them to the next layer to get a final vector.

Keras provides different sorts of layers, such as input layers, dense layers, dropout layers, embedding layers, activation layers, recurrent layers, and so on. Let's get to know some useful layers one by one:

- **Input layer**: The input layer is responsible for sending our input data to the rest of the network. While initializing an input layer, we provide the input data shape.

- **Dense layers**: Dense layers transform the input of a given shape to the output shape we want. Layer 2 in *Figure 8.8* represents a dense layer, which collapses a 5-dimensional input into a 1-dimensional output.

- **Recurrent layers**: Keras provides strong support for RNN, GRU, and LSTM cells. If you're not familiar with RNN variations at all, please refer to the resources in the *Technical requirements* section. We'll use an LSTM layer in our code. The *LSTM layer* subsection contains the input and output shape information. In the next subsection, *Sequential modeling with LSTMs*, we'll get into details of modeling with LSTMs.

- **Dropout layers**: Dropout is a technique to prevent overfitting. Overfitting happens when neural networks memorize data instead of learning it. Dropout layers randomly select a given number of neurons and set their weights to zero for the forward and backward passes, that is, for one iteration. We usually place dropout layers after dense layers.

These are the basic layers that are used in NLP models. The next subsection is devoted to modeling sequential data with LSTMs, which is the core of statistical modeling for NLP.

Sequential modeling with LSTMs

LSTM is an RNN variation. **RNNs** are special neural networks that can process sequential data in steps. In usual neural networks, we assume that all the inputs and outputs are independent of each other. Of course, it's not true for text data. Every word's presence depends on the neighbor words.

For example, during a machine translation task, we predict a word by considering all the words we predicted before. RNNs capture information about the past sequence elements by holding a **memory** (called **hidden state**). *Figure 8.9* shows a well-known illustration of RNNs. The loop on the left-hand side of the figure explains that an RNN feeds the output of the previous step to itself as the current input. The right-hand side of the figure shows the unrolled version of the RNN diagram. At each time step, i, we feed the input word xi, and RNN outputs a value, hi, for this time step:

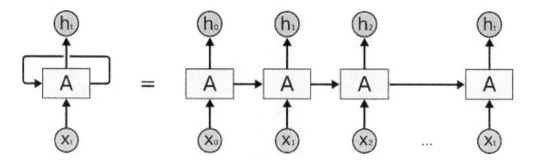

Figure 8.9 – RNN illustration, taken from Colah's notable blog on LSTMs

LSTMs were invented to fix some computational problems of RNNs. RNNs have the problem of forgetting some data back in the sequence, as well as some numerical stability issues due to chain multiplications called **vanishing and exploding gradients**. If you are interested, you can refer to Colah's blog, the link to which you will find in the *References* section.

An LSTM cell is slightly more complicated than an RNN cell, but the logic of computation is the same: we feed one input word at each time step and LSTM outputs an output value at each time step. The following diagram shows what's inside an LSTM cell. Note that the input steps and output steps are identical to the RNN counterparts:

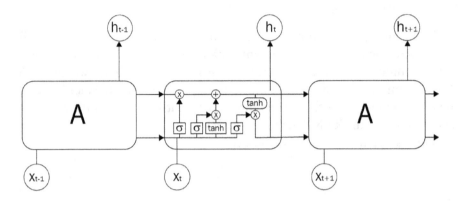

Figure 8.10 – LSTM illustration from Colah's LSTM blog article

Keras has extensive support for the RNN variations GRU and LSTM, as well as a simple API for training RNNs. RNN variations are crucial for NLP tasks, as language data's nature is sequential: text is a sequence of words, speech is a sequence of sounds, and so on.

Now that we have learned what type of statistical model to use in our design, we can switch to a more practical subject: how to represent a sequence of words. In the next section, we'll learn how to transform a sequence of words into a sequence of word IDs and build vocabularies at the same time with Keras's preprocessing module.

Keras Tokenizer

As we remarked in the previous section, text is sequential data (a sequence of words or characters). We'll feed a sentence as a sequence of words. Neural networks can work only with vectors, so we need a way to vectorize the words. In *Chapter 5, Working with Word Vectors and Semantic Similarity*, we saw how to vectorize words with word vectors. A word vector is a continuous representation of a word. In order to vectorize a word, we follow these steps:

1. We tokenize each sentence and turn sentences into a sequence of words.

2. We create a vocabulary from the set of words present in *step 1*. These are words that are supposed to be recognized by our neural network design.

3. Creating a vocabulary should assign an ID to each word.

4. Then word IDs are mapped to word vectors.

Let's look at a short example. We can work on a small corpus of three sentences:

```
data = [
"Tomorrow I will visit the hospital.",
"Yesterday I took a flight to Athens.",
"Sally visited Harry and his dog."
]
```

Let's first tokenize the words into sentences:

```
import spacy
nlp = spacy.load("en_core_web_md")
sentences = [[token.text for token in nlp(sentence)] for
sentence in data]
for sentence in sentences:
    sentence
...
['Tomorrow', 'I', 'will', 'visit', 'the', 'hospital', '.']
['Yesterday', 'I', 'took', 'a', 'flight', 'to', 'Athens', '.']
['Sally', 'visited', 'Harry', 'and', 'his', 'dog', '.']
```

In the preceding code, we iterated over all tokens of the Doc object generated by calling `nlp(sentence)`. Notice that we didn't filter out the punctuation marks. Filtering punctuation depends on the task. For instance, in sentiment analysis, punctuation marks such as "!" correlate to the result. In this example, we'll keep the punctuation marks as well.

Keras's text preprocessing module is used to create vocabularies and trn word sequences into word-ID sequences with the `Tokenizer` class. The following code segment exhibits how to use a `Tokenizer` object:

```
from tensorflow.keras.preprocessing.text import Tokenizer
tokenizer = Tokenizer(lower=True)
tokenizer.fit_on_texts(data)
tokenizer
<keras_preprocessing.text.Tokenizer object at 0x7f89e9d2d9e8>
tokenizer.word_index
{'i': 1, 'tomorrow': 2, 'will': 3, 'visit': 4, 'the': 5,
'hospital': 6, 'yesterday': 7, 'took': 8, 'a': 9, 'flight': 10,
'to': 11, 'athens': 12, 'sally': 13, 'visited': 14, 'harry':
15, 'and': 16, 'his': 17, 'dog': 18}
```

We did the following:

1. We imported `Tokenizer` from the Keras text preprocessing module.

2. We created a `tokenizer` object with the parameter `lower=True`, which means `tokenizer` should lower all words while building the vocabulary.

3. We called `tokenizer.fit_on_texts` on `data` to build the vocabulary. `fit_on_text` works on a sequence of words; input should always be a list of words.

4. We examined the vocabulary by printing `tokenizer.word_index`. `Word_index` is basically a dictionary where keys are vocabulary words and values are word-IDs.

In order to get a word's word-ID, we call `tokenizer.texts_to_sequences`. Notice that the input to this method should always be a list, even if we want to feed only one word. In the following code segment, we feed one-word input as a list (notice the list brackets):

```
tokenizer.texts_to_sequences(["hospital"])
[[6]]
tokenizer.texts_to_sequences(["hospital", "took"])
[[6], [8]]
```

The reverse of `texts_to_sequences` is the `sequences_to_texts`. `sequences_to_texts` method will input a list of lists and return the corresponding word sequences:

```
tokenizer.sequences_to_texts([[3,2,1]])
['will tomorrow i']
tokenizer.sequences_to_texts([[3,2,1], [5,6,10]])
['will tomorrow i', 'the hospital flight']
```

We also notice that the word-IDs start from 1, not 0. 0 is a reserved value and has a special meaning, which means a padding value. Keras cannot process sentences of different lengths, hence we need to pad shorter sentences to reach the longest sentence's length. We pad each sentence of the dataset to a maximum length by adding padding words either to the start or end of the sentence. Keras inserts 0 for the padding, which means it's not a real word, but a padding value. Let's understand padding with a simple example:

```
from tensorflow.keras.preprocessing.sequence import pad_
sequences
sequences = [[7], [8,1], [9,11,12,14]]
```

```
MAX_LEN=4
pad_sequences(sequences, MAX_LEN, padding="post")
array([[ 7,   0,   0,   0],
       [ 8,   1,   0,   0],
       [ 9,  11,  12,  14]], dtype=int32)
pad_sequences(sequences, MAX_LEN, padding="pre")
array([[ 0,   0,   0,   7],
       [ 0,   0,   8,   1],
       [ 9,  11,  12,  14]], dtype=int32)
```

Our sequences are of length 1, 2, and 4. We called pad_sequences on this list of sequences and every sequence is padded with zeros such that its length reaches MAX_LEN=4, the length of the longest sequence. We can pad the sequences from the right or left with the post and pre options. In the preceding code, we padded our sentences with the post option, hence the sentences are padded from the right.

If we put it all together, the complete text preprocessing steps are as follows:

```
from tensorflow.keras.preprocessing.text import Tokenizer
from tensorflow.keras.preprocessing.sequence import pad_sequences
tokenizer = Tokenizer(lower=True)
tokenizer.fit_on_texts(data)
seqs = tokenizer.texts_to_sequences(data)
MAX_LEN=7
padded_seqs = pad_sequences(seqs, MAX_LEN, padding="post")
padded_seqs
array([[ 2,  1,  3,  4,  5,  6,  0],
       [ 7,  1,  8,  9, 10, 11, 12],
       [13, 14, 15, 16, 17, 18,  0]], dtype=int32)
```

Now, we've transformed sentences into a sequence of word-IDs. We've come one step closer to vectorizing the words. In the next subsection, we'll finally transform words into vectors. Then our sentences will be ready to be fed into the neural network.

Embedding words

We're ready to transform words into word vectors. Embedding words into vectors happens via an embedding table. An embedding table is basically a lookup table. Each row holds the word vector of a word. We index the rows by word-IDs, hence the flow of obtaining a word's word vector is as follows:

1. **word->word-ID**: In the previous section, we obtained a word-ID for each word with Keras' `Tokenizer`. `Tokenizer` holds all the vocabulary and maps each vocabulary word to an ID, which is an integer.

2. **word-ID->word vector**: A word-ID is an integer and therefore can be used as an index to the embedding table's rows. Each word-ID corresponds to one row and when we want to get a word's word vector, we first obtain its word-ID and then do a lookup in the embedding table rows with this word-ID.

The following diagram shows how embedding words into word vectors works:

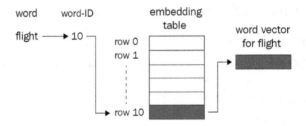

Figure 8.11 – Steps of transforming a word to its word vector with Keras

Remember that in the previous section, we started with a list of sentences. Then we did the following:

1. We broke each sentence into words and built a vocabulary with Keras' `Tokenizer`.

2. The `Tokenizer` object held a word index, which was a **word->word-ID** mapping.

3. After obtaining the word-ID, we could do a lookup to the embedding table rows with this word-ID and got a word vector.

4. Finally, we fed this word vector to the neural network.

Training a neural network is not easy. We have to take several steps to transform sentences into vectors. After these preliminary steps, we're ready to design the neural network architecture and do the model training.

Neural network architecture for text classification

In this section, we will design the neural network architecture for our text classifier. We'll follow these steps to train the classifier:

1. First, we'll preprocess, tokenize, and pad the review sentences. After this step, we'll obtain a list of sequences.

2. We'll feed this list of sequences to the neural network through the input layer.

3. Next, we'll vectorize each word by looking up its word ID in the embedding layer. At this point, a sentence is now a sequence of word vectors, each word vector corresponding to a word.

4. After that, we'll feed the sequence of word vectors to LSTM.

5. Finally, we'll squash the LSTM output with a sigmoid layer to obtain class probabilities.

Let's get started by remembering the dataset again.

Dataset

We'll use the same Amazon fine food reviews dataset from the *Sentiment analysis with spaCy* section. We already processed the dataset with pandas in that section and reduced it to two columns and binary labels. Here is how the `reviews_df` dataset looks:

Out[17]:

	Text	Score
0	To be fair only one of my twins got gas from t...	0
1	Just recieved our first order of these (they d...	1
2	This is a fantastic Chai Masala. I am very pic...	1
3	Flavorful and has added nutrition! You use le...	1
4	I thought this soup tasted good. I liked the t...	0

Figure 8.12 – Result of the reviews_df.head()

We'll transform our dataset a bit. We'll extract the review text and review label from each dataset row and append them into Python lists:

```
train_examples = []
labels = []
for index, row in reviews_df.iterrows():
```

```
text = row["Text"]
rating = row["Score"]
labels.append(rating)
tokens = [token.text for token in nlp(text)]
train_examples.append(tokens)
```

Notice that we appended a list of words to `train_examples`, hence each element of this list is a list of words. Next, we'll invoke Keras' `Tokenizer` on this list of words to build our vocabulary.

Data and vocabulary preparation

We already processed our dataset, hence we are ready to tokenize the dataset sentences and create a vocabulary. Let's go step by step:

1. First, we'll do the necessary imports:

   ```
   from tensorflow.keras.preprocessing.text import Tokenizer
   from tensorflow.keras.preprocessing.sequence import pad_
   sequences
   import numpy as np
   ```

2. We're ready to fit the `Tokenizer` object on our list of words. First, we'll fit the `Tokenizer`, then we'll convert words to their IDs by calling `texts_to_sequences`:

   ```
   tokenizer = Tokenizer(lower=True)
   tokenizer.fit_on_texts(train_examples)
   sequences = tokenizer.texts_to_sequences(train_examples)
   ```

3. Then, we'll pad the short sequences to a maximum length of 50 (we picked this number). Also, this will truncate long reviews to a length of 50 words:

   ```
   MAX_LEN = 50
   X = pad_sequences(sequences, MAX_LEN, padding="post")
   ```

4. Now X is a list of sequences of 50 words. Finally, we'll convert this list of reviews and the labels to numpy arrays:

   ```
   X = np.array(X)
   y = np.array(labels)
   ```

At this point, we're ready to feed our data to our neural network. We'll feed our data to the input layer. For all the necessary imports, please follow the notebook of this section from our GitHub repository: `https://github.com/PacktPublishing/Mastering-spaCy/blob/main/Chapter08/Keras_train.ipynb`.

Here, notice that we didn't do any lemmatization/stemming or stopwords removal. This is completely fine and indeed the standard way to go with neural network algorithms, because words that are variations of the same root word (liked, liking, like) will obtain similar word vectors (recall from *Chapter 5, Working with Word Vectors and Semantic Similarity* that similar words obtain similar word vectors). Also, stopwords occur frequently in different contexts, hence neural network can deduce that these words are just common words of the language and don't carry much importance.

The input layer

The following piece of code defines our input layer:

```
sentence_input = Input(shape=(None,))
```

Don't be confused by None as the input shape. Here, None means that this dimension can be any scalar number, hence, we use this expression when we want Keras to infer the input shape.

The embedding layer

We define the embedding layer as follows:

```
embedding = Embedding(\
input_dim = len(tokenizer.word_index)+1,\
output_dim = 100)(sentence_input)
```

While defining the embedding layer, the input dimension should always be the number of words in the vocabulary (here, there's a plus 1 because the indices start from 1, not 0. Index 0 is reserved for the padding value).

Here, we chose the output shape to be 100, hence the word vectors for the vocabulary words will be 100-dimensional. Popular numbers for word vector dimensions are 50, 100, and 200 depending on the complexity of the task.

The LSTM layer

We'll feed the word vectors to our LSTM:

```
LSTM_layer = LSTM(units=256)(embedding)
```

Here, the units parameter means the dimension of the hidden state. The LSTM output shape and hidden state shape are the same due to the LSTM architecture. Here, our LSTM layer will output a 256-dimensional vector.

The output layer

We obtained a 256-dimensional vector from the LSTM layer and we want to squash it to a 1-dimensional vector (possible values of this vector are 0 and 1, which are the class labels):

```
output_dense = Dense(1, activation='sigmoid')(LSTM_layer)
```

We used the sigmoid function to squash the values. The sigmoid function is an S-shaped function and maps its input to a [0-1] range. You can find out more about this function at https://deepai.org/machine-learning-glossary-and-terms/sigmoid-function.

Compiling the model

After defining the model, we need to compile it with an optimizer, a loss function, and an evaluation metric:

```
model = \
Model(inputs=[sentence_input],outputs=[output_dense])
model.compile(optimizer="adam",loss="binary_crossentropy",\
metrics=["accuracy"])
```

Adaptive Moment Estimation (ADAM) is a popular optimizer in deep learning. It basically adapts how fast the neural network should learn. You can learn about different optimizers in this blog post: https://ruder.io/optimizing-gradient-descent/. Binary cross-entropy is a loss that is used in binary classification tasks. Keras supports different loss functions depending on the tasks. You can find the list on the Keras website at https://keras.io/api/losses/.

A **metric** is a function that we use to evaluate our model's performance. The accuracy metric basically compares how many times the predicted label and the real label matches. A list of supported metrics can be found in Keras's documentation (https://keras.io/api/metrics/).

Fitting the model and experiment evaluation

Finally, we'll fit the model on our data:

```
model.fit(x=X,
          y=y,
          batch_size=64,
          epochs=5,
          validation_split=0.2)
```

Here, x is the list of training examples and y is the list of labels. We want to make 5 passes over the data, hence we set the epochs parameter to 5.

We went over the data 5 times in batch sizes of 64. Usually, we don't fit all of the dataset into the memory at once (due to memory limitations), but we feed the dataset to the classifier in smaller chunks, each chunk being called a **batch**. Here, the parameter batch_size=64 means we want to feed a batch of 64 training sentences at once.

Finally, the parameter validation_split is used to evaluate the experiment. This parameter simply will separate 20 percent of the data as the validation set and validate the model on this validation set. Our experiment results in 0.795 accuracy, which is quite good for such a basic neural network design.

We encourage you to experiment more. You can experiment with the code more by placing dropout layers at different locations (such as after the embedding layer or after the LSTM layer). Another way of experimenting is to try different values for the embedding dimensions, such as 50, 150, and 200, and observe the change in the accuracy. The same applies to the LSTM layer's hidden dimension – you can experiment with different values instead of 256.

We finished training with tf.keras in this section and also concluded the chapter. Keras is a great, efficient, and user-friendly deep learning API; the spaCy and Keras combination is especially powerful. Text classification is an essential task of NLP and we discovered how to do this task with spaCy.

Summary

We have finished this chapter about a very hot NLP topic – text classification. In this chapter, you first learned about text classification concepts such as binary classification, multilabel classification, and multiclass classification. Next, you learned how to train TextCategorizer, spaCy's text classifier component. You learned how to transform your data into spaCy training format and then train the TextCategorizer component with this data.

After learning text classification with spaCy's `TextCategorizer`, in the final section, you learned how to combine spaCy code and Keras code. First, you learned the basics of neural networks, including some handy layers such as the dense layer, dropout layer, embedding layer, and recurrent layers. Then, you learned how to tokenize and preprocess the data with Keras' `Tokenizer`.

You had a quick review of sequential modeling with LSTMs, as well as recalling word vectors from *Chapter 5, Working with Word Vectors and Semantic Similarity*, to understand the embedding layer better. Finally, you went through neural network design with `tf.keras` code. You learned how to design and evaluate a statistical experiment with LSTM.

Looks like a lot! Indeed, it is a lot of material; no worries if it takes time to digest. Practicing text classification can be intense, but in the end, you earn crucial NLP skills.

The next chapter is again devoted to a brand-new technology: **transformers**. In the next chapter, we'll explore how to design high-accuracy NLP pipelines in only a few lines. Let's move onto the next chapter and see what transformers offer for your NLP skills!

References

It'd be good but not mandatory if you're familiar with neural networks, particularly RNN variations. Here is some great material for neural networks:

- Free online book: *Neural Networks and Deep Learning* (`http://neuralnetworksanddeeplearning.com/`)

- Video tutorial at `https://www.youtube.com/watch?v=ob1yS9g-Zcs`

RNN variations, especially LSTMs, have great tutorials too:

- RNN tutorial on the WildML blog: `http://www.wildml.com/2015/09/recurrent-neural-networks-tutorial-part-1-introduction-to-rnns/`

- RNN tutorial by the University of Toronto: `https://www.cs.toronto.edu/~tingwuwang/rnn_tutorial.pdf`

- Colah's blog: `https://colah.github.io/posts/2015-08-Understanding-LSTMs/`

- Blog post by Michael Phi: `https://towardsdatascience.com/illustrated-guide-to-lstms-and-gru-s-a-step-by-step-explanation-44e9eb85bf21`

- Video tutorial at `https://www.youtube.com/watch?v=1WkFhVq9-nc`

Although we have introduced neural networks in this chapter, you can read these references in order to learn more about how neural networks work. More explanations on neural network and LSTM concepts will follow in *Chapter 10, Putting Everything Together: Designing Your Chatbot with spaCy.*

9
spaCy and Transformers

In this chapter, you will learn about the latest hot topic in NLP, transformers, and how to use them with TensorFlow and spaCy.

First, you will learn about transformers and transfer learning. Second, you'll learn about the architecture details of the commonly used Transformer architecture – **Bidirectional Encoder Representations from Transformers** (**BERT**). You'll also learn how **BERT Tokenizer** and **WordPiece** algorithms work. Then you will learn how to quickly get started with pre-trained transformer models of the **HuggingFace** library. Next, you'll practice how to fine-tune HuggingFace Transformers with TensorFlow and Keras. Finally, you'll learn how *spaCy v3.0* integrates transformer models as pre-trained pipelines.

By the end of this chapter, you will be completing the statistical NLP topics of this book. You will add your knowledge of transformers to the knowledge of Keras and TensorFlow that you acquired in *Chapter 8, Text Classification with spaCy*. You'll be able to build state-of-the-art NLP models with just a few lines of code with the power of Transformer models and transfer learning.

In this chapter, we're going to cover the following main topics:

- Transformers and transfer learning
- Understanding BERT
- Transformers and TensorFlow
- Transformers and spaCy

Technical requirements

In this chapter, we'll use the `transformers` and `tensorflow` Python libraries along with `spaCy`. You can install these libraries via `pip`:

```
pip install transformers
pip install "tensorflow>=2.0.0"
```

The chapter code can be found at the book's GitHub repository: `https://github.com/PacktPublishing/Mastering-spaCy/blob/main/Chapter09`.

Transformers and transfer learning

A milestone in NLP happened in 2017 with the release of the research paper *Attention Is All You Need*, by Vaswani et al. (`https://arxiv.org/abs/1706.03762`), which introduced a brand-new machine learning idea and architecture – transformers. Transformers in NLP is a fresh idea that aims to solve sequential modeling tasks and targets some problems introduced by **long short-term memory** (**LSTM**) architecture (recall LSTM architecture from *Chapter 8, Text Classification with spaCy*). Here's how the paper explains how transformers work:

> *"The Transformer is the first transduction model relying entirely on self-attention to compute representations of its input and output without using sequence-aligned RNNs or convolution."*

Transduction in this context means transforming input words to output words by transforming input words and sentences into vectors. Typically, a transformer is trained on a huge corpus such as Wiki or news. Then, in our downstream tasks, we use these vectors as they carry information regarding the word semantics, sentence structure, and sentence semantics (we'll see how to use the vectors precisely in our code in the *Transformers and TensorFlow* section).

We already explored the idea of pre-trained word vectors in *Chapter 5, Working with Word Vectors and Semantic Similarity*. Word vectors such as Glove and FastText vectors are already trained on the Wikipedia corpus and we used them directly for our semantic similarity calculations. In this way, we imported information about word semantics from the Wiki corpus into our semantic similarity calculations. Importing knowledge from pre-trained word vectors or pre-trained statistical models is called **transfer learning**.

Transformers offer thousands of pre-trained models to perform NLP tasks, such as text classification, text summarization, question answering, machine translation, and natural language generation in more than 100 languages. Transformers aim to make state-of-the-art NLP accessible to everyone.

The following screenshot shows a list of the Transformer models provided by HuggingFace (we'll learn about HuggingFace Transformers in the *HuggingFace Transformers* section). Each model is named with a combination of the architecture name (*Bert, DistilBert*, and so on), possibly the language code (*en, de, multilingual*, and similar, given on the left side of the following screenshot), and information regarding whether the model is cased or uncased (the model distinguishes between uppercase and lowercase characters).

Also, on the left-hand side of *Figure 9.1*, we see the task names. Each model is labeled with a task name. We select a model that is suitable for our task, such as text classification or machine translation:

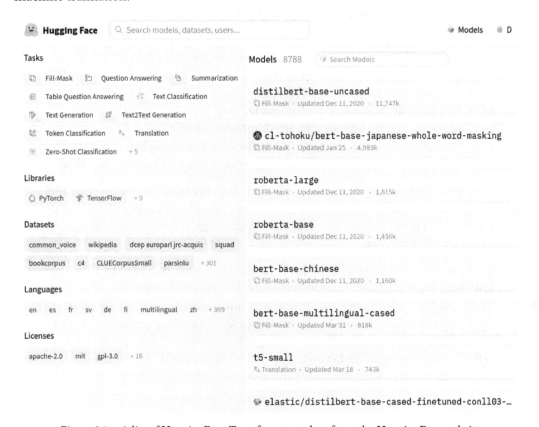

Figure 9.1 – A list of HuggingFace Transformers, taken from the HuggingFace website

To understand what's great about transformers, we'll first revisit LSTM architecture from *Chapter 8, Text Classification with spaCy*. In the previous chapter, we already stepped into the statistical modeling world with Keras and LSTM architecture. LSTMs are great for modeling text; however, they have some shortcomings too:

- LSTM architecture sometimes has difficulties with learning long text. Statistical dependencies in a long text can be difficult to represent by an LSTM because, as the time steps pass, LSTM can forget some of the words that were processed at earlier time steps.

- The nature of LSTMs is sequential. We process one word at each time step. Obviously, parallelizing the learning process is not possible; we have to process sequentially. Not allowing parallelization creates a performance bottleneck.

Transformers address these problems by not using recurrent layers at all. If we have a look at the following, the architecture looks completely different from an LSTM architecture. Transformer architecture consists of two parts – an input encoder (called the **Encoder**) block on the left, and the output decoder (called the **Decoder**) block on the right. The following diagram is taken from this paper and exhibits the transformer architecture:

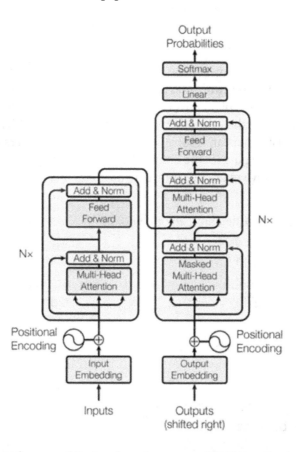

Figure 9.2 – Transformer architecture from the paper entitled "Attention is All You Need"

The preceding architecture is invented for a machine translation task; hence, the input is a sequence of words from the source language, and the output is a sequence of words in the target language. The encoder generates a vector representation of the input words and passes them to the decoder (the word vector transfer is represented by the arrow from the encoder block in the direction of the decoder block). The decoder takes these input word vectors, transforms the output words into word vectors, and finally generates the probability of each output word (labeled in *Figure 9.2* as **Output Probabilities**).

Inside the encoder and decoder blocks, we see feedforward layers, which are basically a dense layer we used in *Chapter 8, Text Classification with spaCy*. The innovation transformers bring lies in the **Multi-Head Attention** block. This block creates a dense representation for each word by using a self-attention mechanism. The **Self-attention** mechanism relates each word in the input sentence to the other words in the input sentence. The word embedding of each word is calculated by taking a weighted average of the other words' embeddings. This way, the importance of each word in the input sentence is calculated, so the architecture focuses its *attention* on each input word in turn.

The following diagram is taken from the original paper and illustrates self-attention. The diagram illustrates how the input words on the left-hand side attend the input word "**it**" on the right-hand side. Darker colors mean more relevance, hence the words "**the animal**" are more related to "**it**" rather than the other words in this sentence. What does this mean? This means that the transformer can resolve what the pronoun "**it**" refers to precisely in this sentence, which is the phrase "**the animal**." This is a great accomplishment of transformers; they can resolve many semantic dependencies in a given sentence:

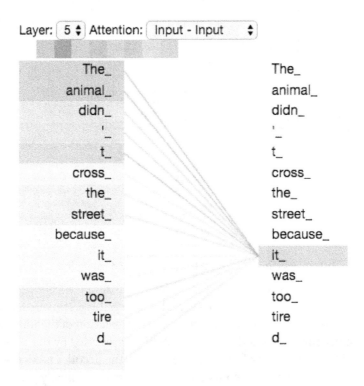

Figure 9.3 – Illustration of the self-attention mechanism

If you want to learn about the details of the Transformer architecture, you can visit `http://jalammar.github.io/illustrated-transformer/`. This talk on YouTube also explains the self-attention mechanism and transformers for all levels of NLP developers: `https://www.youtube.com/watch?v=rBCqOTEfxvg`.

We have already seen that there is a variety of transformer architectures and, depending on the task, we use different types of transformers for different tasks, such as text classification and machine translation. In the rest of this chapter, we'll work with a very popular transformer architecture – BERT. Let's see the BERT architecture and how to use it in our NLP applications in the next section.

Understanding BERT

In this section, we'll explore the most influential and commonly used Transformer model, BERT. BERT is introduced in Google's research paper here: `https://arxiv.org/pdf/1810.04805.pdf`.

What does BERT do exactly? To understand what BERT outputs, let's dissect the name:

- **Bidirectional**: Training on the text data is bi-directional, which means each input sentence is processed from left to right as well as from right to left.
- **Encoder**: An encoder encodes the input sentence.
- **Representations**: A representation is a word vector.
- **Transformers**: The architecture is transformer-based.

BERT is essentially a trained transformer encoder stack. Input into BERT is a sentence, and the output is a sequence of word vectors. The word vectors are contextual, which means that a word vector is assigned to a word based on the input sentence. In short, BERT outputs **contextual word representations**.

We have already seen a number of issues that transformers aim to solve in the previous section. Another problem that transformers address concerns word vectors. In *Chapter 5, Working with Word Vectors and Semantic Similarity*, we saw that word vectors are context-free; the word vector for a word is *always* the same independent of the sentence it is used in. The following diagram explains this problem:

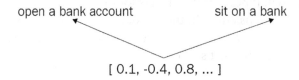

Figure 9.4 – Word vector for the word "bank"

Here, even though the word **bank** has two completely different meanings in these two sentences, the word vectors are the same, because Glove and FastText are **static**. Each word has only one vector and vectors are saved to a file following training. Then, we download these pre-trained vectors and load them into our application.

On the contrary, BERT word vectors are *dynamic*. BERT can generate different word vectors for the same word depending on the input sentence. The following diagram shows the word vectors generated by BERT, in contrast to the word vector in *Figure 9.4*:

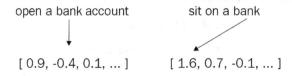

Figure 9.5 – Two distinct word vectors generated by BERT for the same word, "bank," in two different contexts

How does BERT generate these word vectors? In the next section, we'll explore the details of the BERT architecture.

BERT architecture

As has already been remarked in the previous section, BERT is a transformer encoder stack, which means that several encoder layers are stacked on top of each other. The first layer initializes the word vectors randomly, and then each encoder layer transforms the output of the previous encoder layer. The paper introduces two model sizes for BERT: BERT Base and BERT Large. The following diagram shows the BERT architecture:

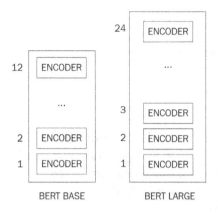

Figure 9.6 – BERT Base and Large architectures, having 12 and 24 encoder layers, respectively

Both BERT models have a huge number of encoder layers. BERT Base has 12 encoder layers and BERT Large has 24 encoder layers. The dimensions of the resulting word vectors are different too; BERT Base generates word vectors of size 768 and BERT Large generates word vectors of size 1024.

As we remarked in the previous section, BERT outputs word vectors for each input word. The following diagram exhibits a high-level overview of BERT inputs and outputs (discard the CLS token for now; you'll learn about it in the *BERT input format* section):

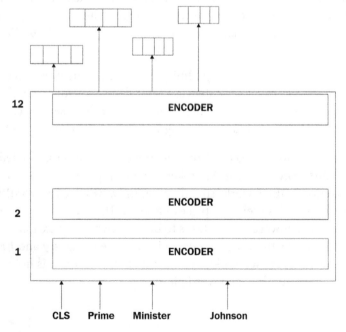

Figure 9.7 – BERT model input word and output word vectors

In the preceding diagram, we can see a high-level overview of BERT inputs and outputs. Indeed, BERT input has to be in a special format and includes some special tokens, such as CLS, in *Figure 9.7*. In the next section, you'll learn about the details of the BERT input format.

BERT input format

We have covered BERT architecture, so let's now understand how to generate the output vectors using BERT. For this purpose, we'll get to know the BERT input data format. The BERT input format can represent a single sentence, as well as a pair of sentences (for tasks such as question answering and semantic similarity, we input two sentences to the model) in a single sequence of tokens.

BERT works with a class of **special tokens** and a special tokenization algorithm called **WordPiece**. Let's get to know the special tokens first. The main special tokens are [CLS], [SEP], and [PAD]:

- The first special token of BERT is [CLS]. The first token of every input sequence has to be [CLS]. We use this token in classification tasks as an aggregate of the input sentence. We ignore this token in non-classification tasks.

- [SEP] means a **sentence separator**. If the input is a single sentence, we place this token at the end of the sentence. If the input is two sentences, then we use this token to separate two sentences. Hence, for a single sentence, the input looks like [CLS] sentence [SEP], and for two sentences, the input looks like [CLS] sentence1 [SEP] sentence2 [SEP].

- [PAD] is a special token meaning **padding**. Recall from the previous chapter that we use padding values to make sentences in our dataset of equal length. BERT receives sentences of a fixed length; hence, we pad the short sentences before feeding them to BERT. The maximum length of tokens we can feed to BERT is **512**.

How about tokenizing the words? Recall from the previous section that we fed a sentence to our Keras model one word at a time. We tokenized our input sentences into words using the spaCy tokenizer. BERT works slightly differently, BERT uses WordPiece tokenization. A "word piece" is literally a piece of a word. The WordPiece algorithm breaks words down into several subwords. The idea is to break down complex/long tokens into simpler tokens. For example, the word playing is tokenized as play and ##ing. A ## character is placed before every word piece to indicate that this token is not a word from the language's vocabulary but that it's a word piece.

Let's take a look at some more examples:

```
playing   play, ##ing
played    play, ##ed
going     go, ##ing
vocabulary = [play,go, ##ing, ##ed]
```

This way, we represent the language vocabulary more compactly as WordPiece groups common subwords. WordPiece tokenization creates wonders on rare/unseen words, as these words are broken down into their subwords.

After tokenizing the input sentence and adding the special tokens, each token is converted to its ID. After that, as a final step, we feed the sequence of token IDs to BERT.

To summarize, this is how we transform a sentence into BERT input format:

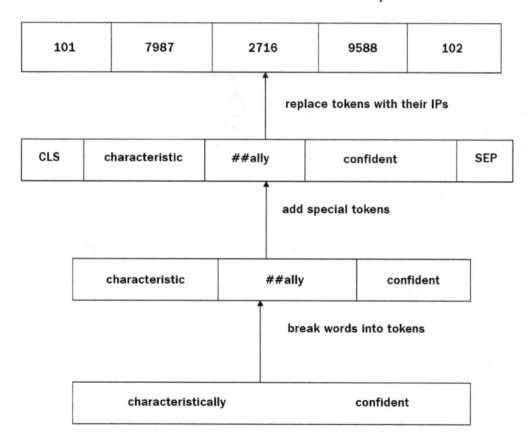

Figure 9.8 – Transforming an input sentence into BERT input format

BERT Tokenizer has different methods for performing all the tasks described previously, but it also has an encoding method that combines these steps into a single step. We'll see how to use BERT Tokenizer in detail in the *Transformers and TensorFlow* section. Before that, we'll learn about the algorithms that are used to train BERT.

How is BERT trained?

BERT is trained on a large unlabeled Wiki corpus and a huge book corpus. Creators of BERT stated the following in Google Research's BERT GitHub repository, `https://github.com/google-research/bert`, as follows:

> *"We then train a large model (12-layer to 24-layer Transformer) on a large corpus (Wikipedia + BookCorpus) for a long time (1M update steps), and that's BERT."*

BERT is trained with two training methods, **masked language model** (**MLM**) and **next sentence prediction** (**NSP**). Let's first go over the details of masked language modeling.

Language modeling is the task of predicting the next token given the sequence of previous tokens. For example, given the sequence of words *Yesterday I visited*, a language model can predict the next token as one of the tokens *church, hospital, school*, and so on. Masked language modeling is a bit different. In this approach, we mask a percentage of the tokens randomly by replacing them with a `[MASK]` token and expect MLM to predict the masked words.

The masked language model in BERT is implemented as follows: First, 15 of the input tokens are chosen at random. Then, the following happens:

1.   80% of the tokens chosen are replaced with `[MASK]`.

2.   10% of the tokens chosen are replaced with another token from the vocabulary.

3.   The remaining 10% are left unchanged.

A training example sentence to LMM looks like the following:

```
[CLS] Yesterday I [MASK] my friend at [MASK] house [SEP]
```

Next, we will look into the details of the other algorithm, NSP.

As the name suggests, NSP is the task of predicting the next sentence given an input sentence. In this approach, we feed two sentences to BERT and expect BERT to predict the order of the sentences, more specifically, if the second sentence is the sentence following the first sentence.

Let's make an example input to NSP. We'll feed two sentences separated by the [SEP] token as input:

```
[CLS] A man robbed a [MASK] yesterday [MASK] 8 o'clock [SEP] He
[MASK] the bank with 6 million dollars [SEP]
Label = IsNext
```

In this example, the second sentence can follow the first sentence; hence, the predicted label is IsNext. How about this example:

```
[CLS] Rabbits like to [MASK] carrots and [MASK] leaves [SEP]
[MASK] Schwarzenegger is elected as the governor of [MASK]
[SEP]
Label= NotNext
```

This example pair of sentences generate the NotNext label, as obviously they are not contextually or semantically related.

That's it! We have learned about BERT architecture; we also learned the details of BERT input data format and how BERT is trained. Now, we're ready to dive into TensorFlow code. In the next section, we'll see how to apply what we learned so far in our TensorFlow code.

Transformers and TensorFlow

In this section, we'll dive into transformers code with TensorFlow. Pre-trained transformer models are provided to the developer community as open source by many organizations, including Google (https://github.com/google-research/bert), Facebook (https://github.com/pytorch/fairseq/blob/master/examples/language_model/README.md), and HuggingFace (https://github.com/huggingface/transformers). All the listed organizations offer pre-trained models and nice interfaces to integrate transformers into our Python code. The interfaces are compatible with either PyTorch or Tensorflow or both.

Throughout this chapter, we'll be using HuggingFace's pre-trained transformers and their TensorFlow interface to the transformer models. HuggingFace is an AI company with a focus on NLP and quite devoted to open source. In the next section, we'll take a closer look at what is available in HuggingFace Transformers.

HuggingFace Transformers

In the first section, we'll discover HuggingFace's pre-trained models, the TensorFlow interface for using these models, and HuggingFace model conventions in general. We saw in *Figure 9.1* that HuggingFace offers different sorts of models. Each model is dedicated to a task such as text classification, question answering, and sequence-to-sequence modeling.

The following diagram is taken from the HuggingFace documentation and shows details of the distilbert-base-uncased-distilled-squad model. In the documentation, first, the task is tagged (upper-left corner of the diagram; the *Question Answering* tag), followed by supported deep learning libraries (PyTorch, TensorFlow, TFLite, TFSavedModel for this model), the dataset it trained on (squad, in this instance), the model language (*en* for English), and the license and base model's name (DistilBERT in this case).

Some models are trained with similar algorithms, and so belong to the same model family. By way of an example, the DistilBERT family includes many models, such as distilbert-base-uncased and distilbert-multilingual-cased. Each model name also includes some information, such as casing (the model recognizes uppercase/lowercase differences) or the model language, such as en, de, or multilingual:

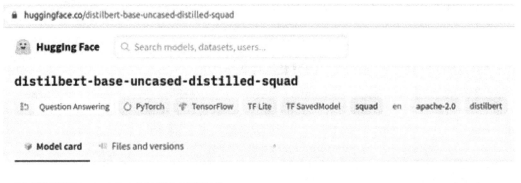

Figure 9.9 – Documentation of the distilbert-base-uncased-distilled-squad model

We already explored BERT in detail in the previous section. HuggingFace documentation provides information about each model family and an individual model's API in detail. *Figure 9.10* shows a list of available models and a list of BERT model architecture variations:

Figure 9.10 – List of the available models on the left-hand side, with a list of the BERT model variations on the right-hand side

The BERT model has many variations for a variety of tasks, such as text classification, question answering, and next sentence prediction. Each of these models is obtained by placing some extra layers on top of the BERT output. Recall from the previous section that the BERT output is a sequence of word vectors for each word of the input sentence. For example, the BERTForSequenceClassification model is obtained by placing a dense layer (we covered dense layers in the previous chapter) on top of the BERT word vectors.

In the rest of this chapter, we'll explore how to use some of these architectures for our tasks as well as how to use BERT word vectors with Keras. Before all of these tasks, we will start with the basic task of tokenization to prepare our input sentences. Let's see the tokenizer code in the next section.

Using the BERT tokenizer

In the *Understanding BERT* section, we already saw that BERT uses the WordPiece algorithm for tokenization. Every input word is broken down into subwords. Let's see how to prepare our input data with the HuggingFace library.

The following lines exhibit the basic usage of the tokenizer:

```
from transformers import BertTokenizer
btokenizer =\
BertTokenizer.from_pretrained('bert-base-uncased')
tokens = btokenizer.tokenize(sentence)
tokens
```

```
['he', 'lived', 'characteristic', '##ally', 'idle', 'and',
'romantic', '.']
ids = btokenizer.convert_tokens_to_ids(tokens)
ids
[2002, 2973, 8281, 3973, 18373, 1998, 6298, 1012]
```

Here are the steps we followed in the preceding code block:

1. First, we imported `BertTokenizer`. Different models have different tokenizers; for instance, XLNet model's tokenizer is called `XLNetTokenizer`.

2. Second, we called the `from_pretrained` method on the tokenizer object and provided the model's name. Note that we don't need to download the pre-trained bert-base-uncased model manually; this method downloads the model by itself.

3. Then, we called the `tokenize` method. `tokenize` basically tokenizes the sentence by breaking all the words down into subwords.

4. We print tokens to examine the subwords. The words "he," "lived," "idle," and so on exist in Tokenizer's vocabulary, and so are kept as they are. "Characteristically" is a rare word, so does not exist in Tokenizer's vocabulary. Then, Tokenizer splits this word into the subwords "characteristic" and "##ally." Notice that "##ally" starts with the characters "##" to emphasize the fact that this is a piece of word.

5. Next, we convert tokens to their token IDs by calling `convert_tokens_to_ids`.

How about [CLS] and [SEP] tokens? In the previous section, we already saw that we have to add these two special tokens to the beginning and end of the input sentence. For the preceding code, we need to involve one more step and add our special tokens manually. Can we do all of these preprocessing steps in a single step, perhaps? The answer is yes; BERT provides a method called `encode` that does the following:

- Adds CLS and SEP tokens to the input sentence

- Tokenizes the sentence by breaking the tokens down into subwords

- Converts the tokens to their token IDs

We call the `encode` method directly on the input sentence as follows:

```
from transformers import BertTokenizer
btokenizer =\
BertTokenizer.from_pretrained('bert-base-uncased')
sentence = "He lived characteristically idle and romantic."
ids = btokenizer.encode(sentence)
```

```
ids
[101, 2002, 2973, 8281, 3973, 18373, 1998, 6298, 1012, 102]
```

This code segment outputs the token IDs in just a single step, instead of calling `tokenize` and `convert_tokens_to_ids` one after the other. The result is a Python list.

How about padding the sentence? We already saw in the previous section that all the input sentences in a dataset should be of equal length because BERT cannot process variable-length sentences. Hence, we need to pad the short sentences to the length of the longest sentence available in the dataset. Also, if we want to use a TensorFlow tensor instead of a plain list, we need to write some conversion code. The HuggingFace library provides `encode_plus` to make our life easier and combine all these steps into one method as follows:

```
from transformers import BertTokenizer
btokenizer =\
BertTokenizer.from_pretrained('bert-base-uncased')
sentence = "He lived characteristically idle and romantic."
encoded = btokenizer.encode_plus(
        text=sentence,
        add_special_tokens=True,
        max_length=12,
        pad_to_max_length=True,
        return_tensors="tf"
)
token_ids = encoded["input_ids"]
print(token_ids)
tf.Tensor([[  101  2002  2973  8281  3973
18373  1998  6298  1012   102     0     0]], shape=(1, 12),
dtype=int32)
```

Here, we called `encode_plus` on our input sentence directly. Our sentence is now padded to a length of 12 (the two 0 IDs at the end of the sequence are the pad tokens), while the special tokens [CLS] and [SEP] are also added to the sentence. The output is directly a TensorFlow tensor, including the token IDs.

The `encode_plus` method takes the following parameters:

- `text`: Input sentence.
- `add_special_tokens`: Add `CLS` and `SEP` tokens.
- `max_length`: The maximum length you want your sentence to be. If the sentence is shorter than `max_length` tokens, we want to pad the sentence.
- `pad_to_max_length`: We feed `True` if we want to pad the sentence, otherwise `False`.
- `return_tensors`: We pass this parameter if we want the output to be a tensor, otherwise the output is a Python list. The available options are `tf` and `pt` for TensorFlow and PyTorch, respectively.

As we can see, BERT Tokenizer provides several methods on input sentences. Preparing data is not so straightforward, but you'll get used to it by practicing. We always encourage you to try out the code examples with your own text.

Now, we're ready to process the transformed input sentences. Let's go ahead and provide our input sentences to the BERT model to obtain BERT word vectors.

> **Tip**
> Always check the name of the tokenizer class that you should use with your transformer. A list of models and their corresponding tokenizers is available at
> `https://huggingface.co/transformers/`.

Obtaining BERT word vectors

In this section, we'll examine the output of the BERT model. As we stated in the *Understanding BERT* section, the output of the BERT model is a sequence of word vectors, one vector per input word. BERT has a special output format and, in this section, we'll examine BERT outputs in detail.

Let's see the code first:

```
from transformers import BertTokenizer, TFBertModel
btokenizer =\
  BertTokenizer.from_pretrained('bert-base-uncased')
bmodel = TFBertModel.from_pretrained("bert-base-uncased")

sentence = "He was idle."
```

```
encoded = btokenizer.encode_plus(
        text=sentence,
        add_special_tokens=True,
        max_length=10,
        pad_to_max_length=True,
        return_attention_mask=True,
        return_tensors="tf"
)
inputs = encoded["input_ids"]
outputs = bmodel(inputs)
```

This code is very similar to the code from the previous section. Here, we also imported `TFBertModel`. After that, we initialized out BERT model with the pre-trained model, `bert-base-uncased`. Then, we transformed our input sentence to BERT input format with `encode_plus` and captured the result, `tf.tensor`, in the input variable. We fed our sentence to the BERT model and captured this output with the `outputs` variable. What's inside the `outputs` variable then?

The output of the BERT model is a tuple of two elements. Let's print the shapes of the output pair:

```
outputs[0].shape
(1, 10, 768)
outputs[1].shape
(1, 768)
```

The first element of the output is the shape `(batch size, sequence length, hidden size)`. We fed only one sentence, hence the batch size here is 1 (the batch size is how many sentences we feed to the model at once). The sequence length here is 10 because we fed `max_length=10` to the tokenizer and padded our sentence to a length of 10. `hidden_size` is a parameter of BERT. In the *BERT Architecture* section, we already remarked that BERT's hidden layer size is 768, and so produces word vectors with a dimension of 768. So, the first output element contains 768-dimensional vectors per word, hence it contains 10 words x 768-dimensional vectors.

The second output is only one vector of 768-dimension. This vector is basically the word embedding of the `[CLS]` token. Recall from the *BERT input format* section that the `[CLS]` token is an aggregate of the whole sentence. You can think of `[CLS]` token's embedding as the pooled version of embeddings of all the words in the sentence. The shape of the second element of the output tuple is always `(batch size, hidden_size)`. Basically, we collect the `[CLS]` token's embedding per input sentence.

Great! We extracted the BERT embeddings. Next, we'll use these embeddings to train our text classification model with TensorFlow and tf.keras.

Using BERT for text classification

In this section, we'll train a binary text classifier with BERT and tf.keras. We'll reuse some of the code from the previous chapter, but this time the code will be much shorter because we'll replace the embedding and LSTM layers with BERT. The complete code is available at the GitHub repository: `https://github.com/PacktPublishing/Mastering-spaCy/blob/main/Chapter09/BERT_spam.ipynb.ipynb`. In this section, we'll skip the data preparation. We used the SMS Spam Collection dataset from Kaggle. You can find the dataset under the `data/` directory at the GitHub repository, too.

Let's get started by importing the BERT models and tokenizer:

```
from transformers import BertTokenizer, TFBertModel
bert_tokenizer =\
BertTokenizer.from_pretrained("bert-base-uncased")
bmodel = TFBertModel.from_pretrained("bert-base-uncased")
```

We have imported the `BertTokenizer` tokenizer and the BERT model, `TFBertModel`. We initialized both the tokenizer and the BERT model with the pre-trained bert-base-uncased model. Notice that the model's name starts with TF – the names of all of the HuggingFace pre-trained models for TensorFlow start with TF. Please pay attention to this detail when you want to play with other transformer models in the future.

We'll also import Keras layers and functions, together with numpy:

```
import numpy as np
import tensorflow
from tensorflow.keras.layers import Dense, Input
from tensorflow.keras.models import Model
```

Now, we're ready to process the input data with `BertTokenizer`:

```
input_ids=[]

for sent in sentences:
    bert_inp=bert_tokenizer.encode_plus(sent,add_special_
tokens = True,max_length =64,pad_to_max_length = True,return_
attention_mask = True)
    input_ids.append(bert_inp['input_ids'])
```

```
input_ids=np.asarray(input_ids)
labels=np.array(labels)
```

As we saw in the *Using the BERT tokenizer* section, this code segment will generate token IDs for each input sentence of the dataset and append them to a list. Labels are the list of class labels and consist of 0 and 1s. Then we convert the Python lists, input_ids, and labels to numpy arrays to feed them to our Keras model.

Finally, we define our Keras model by means of the following lines:

```
inputs = Input(shape=(64,), dtype="int32")
bert = bmodel(inputs)
bert = bert[1]
outputs = Dense(units=1, activation="sigmoid")(bert)
model = Model(inputs, outputs)
```

That's it! We defined our BERT-based text classifier in only five lines of code! Let's dissect the code:

1. First, we defined the input layer, which inputs the sentences to our model. The shape is (64,) because each input sentence is 64 tokens in length. We padded each sentence to the length of 64 tokens when we called the encode_plus method.

2. Next, we fed the input sentences to the BERT model.

3. At the third line, we extracted the second output of the BERT output. Recall from the previous section that the BERT model's output is a tuple. The first element of the output tuple is a sequence of word vectors, and the second element is a single vector that represents the whole sentence, called **pooled output vector**. bert[1] extracts the pooled output vector; this is a vector of shape (1, 768).

4. Next, we squashed the pooled output vector to a vector of shape 1 by a sigmoid function, which is the class label.

5. We defined our Keras model with the inputs and outputs.

Here, the BERT model takes only one line but can transfer the enormous knowledge of the Wiki corpus to your model. At the end of the training, this model obtains an accuracy of 0.96. We usually fit the model for one epoch due to fact that BERT overfits easily even on a moderate size corpus.

The rest of the code handles compiling and fitting the Keras model. Note that BERT has huge memory requirements also. You can see how much RAM is required from Google Research's GitHub link: `https://github.com/google-research/bert#out-of-memory-issues`.

If you have trouble running this section's code on your machine, you can use **Google Colab**, which provides a Jupyter notebook environment through your browser. You can start using Google Colab immediately through `https://colab.research.google.com/notebooks/intro.ipynb`. Our training code runs on Google Colab for around 1.5 hours, while bigger datasets can take more time even though it's just one epoch.

In this section, we learned how to train a Keras model with BERT from scratch. Now, we'll switch to an easier task. We'll explore how to use a pre-trained transformer pipeline. Let's move on to the next section for the details.

Using Transformer pipelines

The HuggingFace Transformers library provides pipelines to help developers benefit from transformer code immediately without any custom training. A **pipeline** is a tokenizer and a pre-trained model combined.

HuggingFace provides a variety of models for a variety of NLP tasks. Here are some tasks that HuggingFace pipelines offer:

- **Sentiment analysis**
- **Question answering**
- **NER**
- **Text summarization**
- **Translation**

You can see the full list of tasks at the Huggingface documentation: `https://huggingface.co/transformers/task_summary.html`. In this section, we'll explore the pipelines for sentiment analysis and question answering (the use of pipelines with other tasks is similar).

Let's go through some examples. We'll start with sentiment analysis:

```
from transformers import pipeline
nlp = pipeline("sentiment-analysis")

sent1 = "I hate you so much right now."
```

```
sent2 = "I love fresh air and exercising."
result1 = nlp(sent1)
result2 = nlp(sent2)
```

In the preceding code snippet, we took the following steps:

1. First, we imported the pipeline function from the `transformers` library. This function creates pipeline objects with the task name given as a parameter. Hence, we created our sentiment analysis pipeline object, `nlp`, by calling this function on the second line.

2. Next, we define two example sentences with negative and positive sentiment, respectively.

3. Then we feed these sentences to the pipeline object, `nlp`.

Here is the output:

```
result1
[{'label': 'NEGATIVE', 'score': 0.9984998}]
result2
[{'label': 'POSITIVE', 'score': 0.99987185}]
```

This worked great! Next, we'll play with question answering. Let's see the code:

```
from transformers import pipeline
nlp = pipeline("question-answering")

res = nlp({
    'question': 'What is the name of this book?',
    'context': "I'll publish my new book Mastering spaCy soon."
})
print(res)
{'score': 0.0007240351873990664, 'start': 25, 'end': 40,
'answer': 'Mastering spaCy'}
```

Again, we imported the pipeline function and used it to create a pipeline object, `nlp`. In *question-answering* tasks, we need to provide a context (the same background information for the model to work on) to the model as well as our question. We asked the model about the name of this book after giving the information that our new publication will be out soon. The answer is `Mastering spaCy`; the transformer worked wonders on this pair! We encourage you to try out your own examples.

We have completed our exploration of HuggingFace transformers. Now, we will move on to our final section of this chapter and see what spaCy offers us as regards transformers.

Transformers and spaCy

spaCy v3.0 was released with great new features and components. The most exciting new feature is undoubtedly **transformer-based pipelines**. The new transformer-based pipelines bring spaCy's accuracy to the state of the art. Integrating transformers into the spaCy NLP pipeline introduced one more pipeline component called **Transformer**. This component allows us to use all HuggingFace models with spaCy pipelines. If we recall from *Chapter 2*, *Core Operations with spaCy*, this is what the spaCy NLP pipeline looks like without transformers:

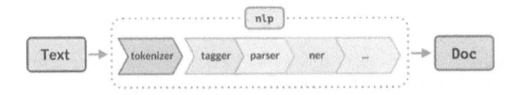

Figure 9.11 – Vector-based spaCy pipeline components

With the release of v3.0, v2 style spaCy models are still supported and transformer-based models are introduced. A transformer-based pipeline component looks like the following:

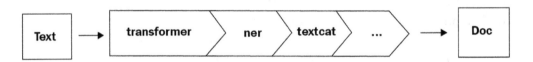

Figure 9.12 – Transformed-based spaCy pipeline components

For each supported language, transformer-based models and v2 style models are listed under the Models page of the documentation (English for an example: `https://spacy.io/models/en`). Transformer-based models can have different size and pipeline components, just like v2 style models. Also, each model has corpus and genre information as well, just like the v2 style models. Here is an example of an English transformer-based language model from the Models page:

en_core_web_trf

English transformer pipeline (roberta-base). Components: transformer, tagger, parser, ner, attribute_ruler, lemmatizer.

LANGUAGE	**EN** English
TYPE	**CORE** Vocabulary, syntax, entities, vectors
GENRE	**WEB** written text (blogs, news, comments)
SIZE	**TRF** 438 MB
COMPONENTS ⑦	transformer, tagger, parser, ner, attribute_ruler, lemmatizer
PIPELINE ⑦	transformer, tagger, parser, ner, attribute_ruler, lemmatizer
VECTORS ⑦	0 keys, 0 unique vectors (0 dimensions)
SOURCES ⑦	OntoNotes 5
AUTHOR	Explosion
LICENSE	MIT

Figure 9.13 – spaCy English transformer-based language models

As we see from the preceding screenshot, the first pipeline component is a transformer and the rest of the pipeline components are the ones we already covered in *Chapter 3, Linguistic Features*. The transformer component generates the word representations and deals with the WordPiece algorithm to tokenize words into subwords. The word vectors are fed to the rest of the pipeline.

Downloading, loading, and using transformer-based models are identical to v2 style models. Currently, English has two pre-trained transformer-based models, en_core_web_trf and en_core_web_lg. Let's get started by downloading the en_core_web_trf model:

```
python3 -m spacy download en_core_web_trf
```

This should produce output similar to the following:

```
Collecting en-core-web-trf==3.0.0
  Downloading https://github.com/explosion/spacy-models/
releases/download/en_core_web_trf-3.0.0/en_core_web_trf-3.0.0-
py3-none-any.whl (459.7 MB)
```

```
|████████████████████████████| 459.7 MB 40 kB/s
Requirement already satisfied: spacy<3.1.0,>=3.0.0 in /usr/
local/lib/python3.6/dist-packages (from en-core-web-trf==3.0.0)
(3.0.5)
```

Once the model download is complete, the following output should be generated:

```
Successfully installed en-core-web-trf-3.0.0 spacy-
alignments-0.8.3 spacy-transformers-1.0.2 tokenizers-0.10.2
transformers-4.5.1
✔ Download and installation successful
You can now load the package via spacy.load('en_core_web_trf')
```

Loading a transformer-based model is identical to what we do for v2 style models, too:

```
import spacy
nlp = spacy.load("en_core_web_trf")
```

After loading our model and initializing the pipeline, we can use this model in the same way we used v2 style models:

```
doc = nlp("I visited my friend Betty at her house.")
doc.ents
(Betty,)
for word in doc:
    print(word.pos_, word.lemma_)
...
PRON I
VERB visit
PRON my
NOUN friend
PROPN Betty
ADP at
PRON her
NOUN house
PUNCT .
```

So far so good, but what's new then? Let's examine some features that come from the transformer component. We can access the features related to the transformer component via doc._.trf_data.trf_data, which contains the word pieces, input ids, and vectors that are generated by the transformer. Let's examine the features one by one:

```
doc = nlp("It went there unwillingly.")
doc._.trf_data.wordpieces
WordpieceBatch(strings=[['<s>', 'It', 'Gwent',
'Gthere', 'Gunw', 'ill', 'ingly', '.', '</s>']],
input_ids=array([[      0,     243,     439,       89,
10963,    1873,    7790,        4,        2]]), attention_mask=array([[1,
1, 1, 1, 1, 1, 1, 1, 1]]), lengths=[9], token_type_ids=None)
```

In the preceding output, we see five elements: word pieces, input IDs, attention masks, lengths, and token type IDs. Word pieces are the subwords that are generated by the WordPiece algorithm. The word pieces of this sentence are as follows:

```
<s>
It
Gwent
Gthere
Gunw
Ill
ingly
.
</s>
```

Here, <s> and </s> are special tokens and are used in the sentence at the beginning and end. The word unwillingly is divided into three subwords – unw, ill, and ingly. The G character is used to mark the word boundaries. The tokens without G are subwords, such as ill and ingly in the preceding word piece list (with the exception of the first word in the sentence, the first word of the sentence is marked by <s>).

Next, we have to take a look at input_ids. Input IDs have the same meaning as the input IDs we introduced in the *Using the BERT tokenizer* section. These are basically the subword IDs assigned by the transformer's tokenizer.

The attention mask is a list of 0s and 1s for pointing the transformer to those tokens it should pay attention to. 0 corresponds to PAD tokens, while all the other tokens should have a corresponding 1.

`lengths` is the length of the sentence after breaking it down into subwords. Here, it's 9 obviously, but notice that `len(doc)` outputs 5, while spaCy always operates on linguistic words.

`token_type_ids` are used by transformer tokenizers to mark the sentence boundaries for two sentence input tasks, such as question and answering. Here, we provide only one text, hence this feature is not applicable.

We can see that the token vectors generated by the transformer, `doc._.trf_data.tensors`, contain the output of the transformer, a sequence of word vectors per word, and the pooled output vector (we introduced these concepts in the *Obtaining BERT word vectors* section. If you need to refresh your memory, please refer to this section):

```
doc._.trf_data.tensors[0].shape
(1, 9, 768)
doc._.trf_data.tensors[1].shape
(1, 768)
```

The first element of the tuple is the vectors for the tokens. Each vector is `768`-dimensional; hence 9 words produce 9 x 768-dimensional vectors. The second element of the tuple is the pooled output vector, which is an aggregate representation for the input sentence, and so is of the shape `1x768`.

This concludes our exploration of spaCy transformer-based pipelines. Once again, we saw that spaCy provides user-friendly API and packaging, even for complicated models such as transformers. Transformer integration is yet another great reason to use spaCy for NLP.

Summary

You have completed an exhaustive chapter about a very hot topic in NLP. Congratulations! In this chapter, you started by learning what sort of models transformers are and what transfer learning is. Then, you learned about the commonly used Transformer architecture, BERT. You learned the architecture details and the specific input format, as well as the BERT Tokenizer and WordPiece algorithm.

Next, you became familiar with BERT code by using the popular HuggingFace Transformers library. You practiced fine-tuning BERT on a custom dataset for a sentiment analysis task with TensorFlow and Keras. You also practiced using pre-trained HuggingFace pipelines for a variety of NLP tasks, such as text classification and question answering. Finally, you explored the spaCy and Transformers integration of the new spaCy release, spaCy v3.0.

By the end of this chapter, you had completed the statistical NLP sections of this book. Now you're ready to put everything you learned together to build a modern NLP pipeline. Let's move on to the next chapter and see how we use our new statistical skills!

10
Putting Everything Together: Designing Your Chatbot with spaCy

In this chapter, you will use everything you have learned so far to design a chatbot. You will perform entity extraction, intent recognition, and context handling. You will use different ways of syntactic and semantic parsing, entity extraction, and text classification.

First, you'll explore the dataset we'll use to collect linguistic information about the utterances within it. Then, you'll perform entity extraction by combining the spaCy **named entity recognition (NER)** model and the spaCy `Matcher` class. After that, you'll perform intent recognition with two different techniques: a pattern-based method and statistical text classification with TensorFlow and Keras. You'll train a character-level LSTM to classify the utterance intents.

The final section is a section dedicated to sentence- and dialog-level semantics. You'll take a deep dive into semantic subjects such as **anaphora resolution**, **grammatical question types**, and **differentiating subjects from objects**.

By the end of this chapter, you'll be ready to design a real chatbot **natural language understanding (NLU)** pipeline. You will bring together what you learned in all previous chapters – linguistically and statistically – by combining several spaCy pipeline components such as **NER**, a **dependency parser**, and a **POS tagger**.

In this chapter, we're going to cover the following main topics:

- Introduction to conversational AI
- Entity extraction
- Intent recognition

Technical requirements

In this chapter, we'll be using NumPy, TensorFlow, and scikit-learn along with spaCy. You can install these libraries via `pip` using the following commands:

```
pip install numpy
pip install tensorflow
pip install scikit-learn
```

You can find the chapter code and data at the book's GitHub repository: `https://github.com/PacktPublishing/Mastering-spaCy/tree/main/Chapter10`.

Introduction to conversational AI

We welcome you to our last and very exciting chapter, where you'll be designing a chatbot NLU pipeline with spaCy and TensorFlow. In this chapter, you'll learn the NLU techniques for extracting meaning from multiturn chatbot-user interactions. By learning and applying these techniques, you'll take a step into **conversational AI development**.

Before diving into the technical details, there's one fundamental question: what is a chatbot? Where can we find one? What exactly does conversational AI mean?

Conversational artificial intelligence (conversational AI) is a field of machine learning that aims to create technology that enables users to have text- or speech-based interactions with machines. Chatbots, virtual assistants, and voice assistants are typical conversational AI products.

A **chatbot** is a software application that is designed to make conversations with humans in chat applications. Chatbots are popular in a wide variety of commercial areas including HR, marketing and sales, banking, and healthcare, as well as in personal, non-commercial areas such as small talk. Many commercial companies, such as Sephora (Sephora owns two chatbots – a virtual make-up artist chatbot on Facebook messenger platform and a customer service chatbot again on Facebook messenger), IKEA (IKEA have a customer service chatbot called Anna), AccuWeather, and many more, own customer service and FAQ chatbots.

Instant messaging services such as Facebook Messenger and Telegram provide interfaces to developers for connecting their bots. These platforms provide detailed guidelines for developers as well, such as the Facebook Messenger API documentation: (`https://developers.facebook.com/docs/messenger-platform/getting-started/quick-start/`) or the Telegram bot API documentation: (`https://core.telegram.org/bots`).

A **virtual assistant** is also a software agent that performs some tasks upon user request or question. A well-known example is **Amazon Alexa**. Alexa is a voice-based virtual assistant and can perform many tasks, including playing music, setting alarms, reading audiobooks, playing podcasts, and giving real-time information for weather, traffic, sports, and so on. Alexa Home can control connected smart home devices and perform a variety of tasks, including switching the lights on and off, controlling the garage door, and so on.

Other well-known examples are Google Assistant and Siri. Siri is integrated into a number of Apple products, including iPhone, iPad, iPod, and macOS. On iPhone, Siri can make calls, answer calls, and send and receive text messages as well as WhatsApp messages. Google Assistant also can perform a wide variety of tasks, such as providing real-time flight, weather, and traffic information; sending and receiving text messages; setting alarms; providing device battery information; checking your email inbox; integrating with smart home devices; and so on. Google Assistant is available on Google Maps, Google Search, and standalone Android and iOS applications.

Here is a list of the most popular and well-known virtual assistants to give you some more ideas of what's out there:

- Amazon Alexa
- AllGenie from Alibaba Group
- Bixby from Samsung
- Celia from Huawei
- Duer from Baidu
- Google Assistant

- Microsoft Cortana

- Siri from Apple

- Xiaowei from Tencent

All of these virtual assistants are voice-based and are usually invoked with a **wake word**. A wake word is a special word or phrase that is used to activate a voice assistant. Some examples are *Hey Alexa*, *Hey Google*, and *Hey Siri*, which are the wake words of Amazon Alexa, Google Assistant, and Siri, respectively. If you want to know more about the development details of these products, please refer to the *References* section of this chapter.

Now, we come to the technical details. What are the NLP components of these products? Let's look at these NLP components in detail.

NLP components of conversational AI products

A typical voice-based conversational AI product consists of the following components:

- **Speech-to-text component**: Converts user speech into text. Input to this component is a WAV/mp3 file and the output is a text file containing the user utterance as a text.

- **Conversational NLU component**: This component performs intent recognition and entity extraction on the user utterance text. The output is the user intent and a list of entities. Resolving references in the current utterance to the previous utterances is done in this component (please refer to the *Anaphora resolution* section).

- **Dialog manager**: Keeps the conversation memory to make a meaningful and coherent chat. You can think of this component as the dialog memory as this component usually holds a **dialog state**. The dialog state is the state of the conversation: the entities that have appeared so far, the intents that have appeared so far, and so on. Input to this component is the previous dialog state and the current user parsed with intent and entities. The output of this component is the new dialog state.

- **Answer generator**: Given all the inputs from the previous stages, generates the system's answer to the user utterance.

- **Text-to-speech**: This component generates a speech file (WAV or mp3) from the system's answer.

Each of the components is trained and evaluated separately. For example, the speech-to-text component is trained on an annotated speech corpus (training is done on speech files and the corresponding transcriptions). The NLU component is trained on intent and an entity labeled corpus (similar to the datasets we used in *Chapters 6, 7, 8, and 9*). In this chapter, we'll focus on the NLU component tasks. For text-based products, the first and last components are not necessary and are replaced with email or chat client integration.

There's another paradigm that is called **end-to-end spoken language understanding (SLU)**. In SLU architectures, the system is trained end to end, which means that the input to the system is a speech file and the output is the system response. Each approach has pros and cons; you can refer to the *References* section for more material.

As the author of this book, I'm happy to present this chapter to you with my domain experience. I've been working in the conversational AI area for quite some time and tackle challenges of language and speech processing every day for our product. Me and my colleagues are building the world's first driver digital assistant, Chris (*Tips & Tricks: How to talk to Chris – basic voice commands*, `https://www.youtube.com/watch?v=Qwnjszu3exY`). Chris can make calls, answer incoming calls, read and write WhatsApp and text messages, play music, navigate, and make small talk. Here is Chris:

Figure 10.1 – In-car voice assistant Chris (this is the product that the author is working on)

As we see from the preceding examples, conversational AI has become a hot topic recently. As an NLP professional, it's quite likely that you'll work for a conversational product or work in a related area such as speech recognition, text-to-speech, or question answering. Techniques presented in this chapter such as intent recognition, entity extraction, and anaphora resolution are applicable to a wide set of NLU problems as well. Let's dive into the technical sections. We'll start by exploring the dataset that we'll use throughout this chapter.

Getting to know the dataset

In *Chapters 6, 7, 8,* and *9,* we worked on well-known real-world datasets for text classification and entity extraction purposes. In these chapters, we always explored our dataset as the very first task. The main point of data exploration is to understand the nature of the dataset text in order to develop strategies in our algorithms that can tackle this dataset. If we recall from *Chapter 6, Putting Everything Together: Semantic Parsing with spaCy,* the following are the main points we should keep an eye on during our exploration:

- What kind of utterances there are? Are utterances short text or full sentences or long paragraphs or documents? What is the average utterance length?

- What sort of entities does the corpus include? Person names, organization names, geographical locations, street names? Which ones do we want to extract?

- How is punctuation used? Is the text correctly punctuated or is no punctuation used at all?

- How are the grammatical rules followed? Is capitalization correct, and did the users follow the grammatical rules? Are there misspelled words?

The previous datasets we used consisted of (`text, class_label`) pairs to be used in text classification tasks or (`text, list_of_entities`) pairs to be used in entity extraction tasks. In this chapter, we'll tackle a much more complicated task, chatbot design. Hence, the dataset will be more structured and more complicated.

Chatbot design datasets are usually in JSON format to maintain the dataset structure. Here, structure means the following:

- Keeping the order of user and system utterances

- Marking slots of the user utterances

- Labeling the intent of the user utterances

Throughout this chapter, we'll use Google Research's **The Schema-Guided Dialogue dataset (SGD)** (`https://github.com/google-research-datasets/dstc8-schema-guided-dialogue`). This dataset consists of annotated user-virtual assistant interactions. The original dataset contains over 20,000 dialog segments in several areas, including restaurant reservations, movie reservations, weather queries, and travel ticket booking. Dialogs include utterances of user and virtual assistant turn by turn. In this chapter, we won't use all of this massive dataset; instead, we'll use a subset about restaurant reservations.

Let's get started with downloading the dataset. You can download the dataset from the book's GitHub repository at `https://github.com/PacktPublishing/Mastering-spaCy/blob/main/Chapter10/data/restaurants.json`. Alternatively, you can write the following code:

```
$ wget https://github.com/PacktPublishing/Mastering-spaCy/blob/
main/Chapter10/data/restaurants.json
```

If you open the file with a text editor and look at the first few lines, you'll see the following:

```
{
    "dialogue_id": "1_00000",
    "turns": [
        {
            "speaker": "USER",
            "utterance": "I am feeling hungry so I would like to find
a place to eat.",
            "slots": [],
            "intent": "FindRestaurants"
        },
        {
            "speaker": "SYSTEM",
            "utterance": "Do you have a specific which you want the
eating place to be located at?",
            "slots": []
        }
```

First of all, the dataset consists of dialog segments and each dialog segment has a `dialogue_id` instance. Each dialog segment is an ordered list of turns and each turn belongs to the user or to the system. A **dialog segment** contains multiple turns; here, the `turns` field is a list of the user/system turns. Each element of the `turns` list is a turn. One turn consists of a speaker (user or system), the speaker's utterance, a list of slots, and an intent for the user utterances.

Here are some example user utterances from the dataset:

```
Hi. I'd like to find a place to eat.
I want some ramen, I'm really craving it. Can you find me an
afforadable place in Morgan Hill?
I would like for it to be in San Jose.
Yes, please make a reservation for me.
No, Thanks
Hi i need a help, i am very hungry, I am looking for a
restaurant
Yes, on the 7th for four people.
No. Can you change it to 1 pm on the 9th?
Yes. What is the phone number? Can I buy alcohol there?
```

As we see from these example utterances, capital letters and punctuation are used in the user utterances. Users can make typos, such as the word `afforadable` in the second sentence. There are some grammatical errors as well, such as the wrong usage of a capital letter in the word `Thanks` of the fifth sentence. Another capitalization mistake occurs in the sixth sentence, where the pronoun *I* is written as i twice.

Also, one utterance can contain multiple sentences. The first utterance starts with a greeting sentence and the last two sentences start with an affirmative or negative answer sentence each. The fourth sentence also starts with a `Yes`, but not as a standalone sentence; instead it's separated from the second sentence with a comma.

Intent recognition for multiple sentence utterances is a point we need to pay attention to in general – these types of utterances can contain multiple intents. Also, answer generation for multi-sentence utterances is a bit tricky; sometimes we need to generate only one answer (such as for the second sentence in the preceding code) or sometimes we need to generate an answer per each user sentence (such as for the last sentence in the preceding code).

This is a dataset for restaurant reservations, so naturally it includes some slots in user utterances such as the location, cuisine, time, date, number of people, and so on. Our dataset includes the following slots:

```
city
cuisine
date
phone_number
restaurant_name
```

```
street_address
time
```

Here are some example sentences with the preceding slot types and their values:

```
Find me Ethiopian/cuisine cuisine in Berkeley/city.
The phone number is 707-421-0835/phone_number. Your reservation
is confirmed.
No, change the time to 7 pm/time and for one person only.
No, change it on next friday/date.
```

Now, we come to the class labels for the intent recognition and the distribution of these class labels. Here's the class labels distribution:

```
552 FindRestaurants
625 ReserveRestaurant
56  NONE
```

NONE is a special class label for utterances that indicate the end of a conversation or just saying thank you. This class of utterances is not related to restaurant reservation in general. Utterances that intend to list restaurants and get some information are labeled with the class label FindRestaurants, and utterances that include the intent to make a booking are labeled with ReserveRestaurants. Let's see some example utterances of each class:

```
No, Thanks  NONE
No, thank you very much. NONE
Nothing much. I'm good.  NONE
I am feeling hungry so I would like to find a place to eat.
FindRestaurants
Hi i need a help, i am very hungry, I am looking for a
restaurant  FindRestaurants
Ok, What is the address? How pricey are they? FindRestaurants
Please can you make the reservation ReserveRestaurant
That's good. Do they serve liquor and what is there number?
ReserveRestaurant
Thank you so much for setting that up. ReserveRestaurant
```

We notice that the follow-up sentences, such as utterances 6, 8, and 9, are marked with the intents `FindRestaurants` and `ReserveRestaurant`. These utterances don't contain the intents of finding/reserving directly, but they continue the dialog about finding/reserving a restaurant and still make queries about the restaurant/reservation. Hence, although there are no explicit actions of finding/reserving stated in these utterances, still the intents are to find/reserve a restaurant.

That's it – we collected enough insights about our dataset using the preliminary work of this section. With these insights, we're ready to build our NLU pipeline. We'll start with extracting the user utterance entities.

Entity extraction

In this section, we'll implement the first step of our chatbot NLU pipeline and extract entities from the dataset utterances. The following are the entities marked in our dataset:

```
city
date
time
phone_number
cuisine
restaurant_name
street_address
```

To extract the entities, we'll use the spaCy NER model and the spaCy `Matcher` class. Let's get started by extracting the `city` entities.

Extracting city entities

We'll first extract the `city` entities. We'll get started by recalling some information about the spaCy NER model and entity labels from *Chapter 3, Linguistic Features*, and *Chapter 6, Putting Everything Together: Semantic Parsing with spaCy*:

- First, we recall that the spaCy named entity label for cities and countries is GPE. Let's ask spaCy to explain what GPE label corresponds to once again:

```
import spacy
nlp = spacy.load("en_core_web_md")
spacy.explain("GPE")
'Countries, cities, states'
```

- Secondly, we also recall that we can access entities of a Doc object via the ents property. We can find all entities in an utterance that are labeled by the spaCy NER model as follows:

```
import spacy
nlp = spacy.load("en_core_web_md")
doc = nlp("Can you please confirm that you want to book
a table for 2 at 11:30 am at the Bird restaurant in Palo
Alto for today")
doc.ents
(2, 11:30 am, Bird, Palo Alto, today)
for ent in doc.ents:
    print(ent.text, ent.label_)
2 CARDINAL
11:30 am TIME
Bird PRODUCT
Palo Alto GPE
today DATE
```

In this code segment, we listed all named entities of this utterance by calling doc.ents. Then, we examined the entity labels by calling ent.label_. Examining the output, we see that this utterance contains five entities – one cardinal number entity (2), one TIME entity (11:30 am), one PRODUCT entity (Bird, which is not an ideal label for a restaurant), one CITY entity (Palo Alto), and one DATE entity (today). The GPE type entity is what we're looking for; Palo Alto is a city in the US and hence is labeled by the spaCy NER model as GPE.

The script at https://github.com/PacktPublishing/Mastering-spaCy/blob/main/Chapter10/extract_city_ents.py in the book's GitHub outputs all the utterances that include a city entity together with the city entities. From the output of this script, we can see that the spaCy NER model performs very well on this corpus for GPE entities. We don't need to train the spaCy NER model with our custom data.

We extracted city entities, and our chatbot knows in which city to look for a restaurant. Now, we'll extract dates and times to allow our chatbot to make a real reservation.

Extracting date and time entities

Extracting DATE and TIME entities is similar to extracting CITY entities, which we saw in the previous section. We'll again go over the corpus utterances and see how successful the spaCy NER model is at extracting DATE and TIME entities from our corpus.

Let's see some example utterances from the corpus:

```
import spacy
nlp = spacy.load("en_core_web_md")
sentences = [
    "I will be eating there at 11:30 am so make it for then.",
    "I'll reach there at 1:30 pm.",
    "No, change it on next friday",
    "Sure. Please confirm that the date is now next Friday and
for 1 person.",
    "I need to make it on Monday next week at half past 12 in
the afternoon.",
    "A quarter past 5 in the evening, please."
]
```

In the following code, we'll extract the entities of these example utterances:

```
for sent in sentences:
    doc = nlp(sent)
    ents = doc.ents
    print([(ent.text, ent.label_) for ent in ents])
[('11:30 am', 'TIME')]
[('1:30 pm', 'TIME')]
[('next friday', 'DATE')]
[('next Friday', 'DATE'), ('1', 'CARDINAL')]
[('Monday next week', 'DATE'), ('half past 12', 'DATE')]
[('A quarter past 5', 'DATE')]
[('the evening', 'TIME'), ('4:45', 'TIME')]
```

Looks good! The output looks quite successful:

- The time entities 11:30 am and 1:30 pm of the first and second sentences are extracted successfully.

- The DATE entities next friday and next Friday of the third and fourth sentences are extracted as well. Notice the first entity includes a typo: friday should be written as *Friday* – still, the spaCy NER model successfully extracted this entity.

- The fifth sentence included both a DATE entity and a TIME entity. We can break the DATE entity Monday next week into two parts: Monday – a weekday and next week – a relative date (the exact date depends on the date of the utterance). This entity consists of two noun phrases: Monday (noun) and next week (adjective noun). spaCy can handle such multiword entities. The time entity, half past 12, of this utterance is also a multiword entity. This entity consists of a noun (half), a preposition (past), and a number (12).

- The same goes for the sixth utterance's multiword TIME entity, A quarter past 5. Here is the dependency tree of this entity:

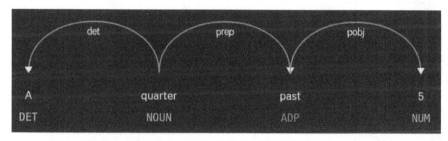

Figure 10.2 – Dependency tree of the time entity "A quarter past 5"

The preceding examples look quite good indeed, but how about the following utterances:

```
sentences = [
    "Have a great day.",
    "Have a nice day.",
    "Have a good day",
    "Have a wonderful day.",
    "Have a sunny and nice day"
]
for sent in sentences:
    doc = nlp(sent)
    ents = doc.ents
    print([(ent.text, ent.label_) for ent in ents])
[('a great day', 'DATE')]
[('a nice day', 'DATE')]
[]
```

```
[]
```
```
[]
```

Oops-a-daisy – looks like we have some **false positives** here. The spaCy NER model labeled some phrases, including the word day, as date entities incorrectly. What can we do here?

Fortunately, these false matches don't form a pattern such as **determiner adjective day**, because the word sequences a good day and a wonderful day of the third and fourth sentence are not labeled as entities. Only the word sequences a great day and a nice day are labeled as entities. Then, we can just filter the spaCy NER results with the following two patterns:

```
sentence = 'Have a nice day.'
doc = nlp(sentence)
wrong_matches = ["a great day", "a nice day"]
date_ents = [ent for ent in doc.ents if ent.label_ == "DATE"]
date_ents = list(filter(lambda e: e.text not in wrong_matches,
date_ents))
date_ents
[]
```

The preceding code block performs the following steps:

1. First, we defined a list of phrases that we don't want to come up as DATE entities.

2. We extracted the DATE entities of the Doc object on the third line by iterating over all entities of doc and picking the entities whose labels were DATE.

3. In the next line, we filtered the entities that didn't appear in the wrong_matches list.

4. We printed the result. As expected, the final result of the date entity is an empty list.

Great, we have extracted DATE and TIME entities along with CITY entities. For all the three entity types, we used the spaCy NER model directly, because spaCy NER recognizes date, time, and location entities. How about phone_number entities? SpaCy NER doesn't include such a label at all. So, we'll use some Matcher class tricks to handle this entity type. Let's extract the phone numbers.

Extracting phone numbers

We had some `Matcher` class practice on entities that include numbers in *Chapter 4, Rule-Based Matching*. We can also recall from *Chapter 4, Rule-Based Matching*, that matching number type entities can be indeed quite tricky; extracting telephone numbers especially requires attention. Phone numbers can come in different formats, with dashes (212-44-44), area codes ((312) 790 12 31), country and area codes (+49 30 456 222), and the number of digits differing from country to country. As a result, we usually examine the following points:

- How many country formats are the corpus phone number entities written in?
- How are the digit blocks separated – with a dash, or whitespace, or both?
- Is there an area code block in some phone numbers?
- Is there a country code block in some phone numbers?
- Are the country code blocks preceded with a + or 00, or are both formats used?

Let's examine some of our phone number entities, then:

> You can call them at 415-775-1800. And they do not serve alcohol.

> Their phone number is 408-374-3400 and they don't have live music.

> Unfortunately no, they do not have live music, however here is the number: 510-558-8367.

All the phone-type entities occur in system utterances. The chatbot fetches phone numbers of restaurants and provides them to the users. The chatbot formed phone number entities by placing a dash between the digit blocks. Also, all the phone numbers are in USA phone number format. Hence the phone number format is uniform and is of the form `ddd-ddd-dddd`. This is very good for defining a Matcher pattern. We can define only one pattern and it matches all the phone number entities.

Let's first see how an example phone number tokenizes:

```
doc= nlp("The phone number is 707-766-7600.")
[token for token in doc]
[The, phone, number, is, 707, -, 766, -, 7600, .]
```

Each digit block is tokenized as one token and each dash character is tokenized as one token as well. Hence, in our Matcher pattern, we'll look for a sequence of five tokens: a three-digit number, a dash, a three-digit number again, a dash again, and finally a four-digit number. Then, our Matcher pattern should look like this:

```
{"SHAPE": "ddd"}, {"TEXT": "-"}, {"SHAPE": "ddd"}, {"TEXT":
"-"}, {"SHAPE": "dddd"}
```

If you recall from *Chapter 4, Rule-Based Matching*, the SHAPE attribute refers to the token shape. The token shape represents the shape of the characters: d means a digit, X means a capital character, and x means a lowercase character. Hence {"SHAPE": "ddd"} means a token that consists of three digits. This pattern will match five tokens of the form ddd-ddd-dddd. Let's try our brand-new pattern on a corpus utterance:

```
from spacy.matcher import Matcher
matcher = Matcher(nlp.vocab)
pattern = [{"SHAPE": "ddd"}, {"TEXT": "-"}, {"SHAPE": "ddd"},
{"TEXT": "-"}, {"SHAPE": "dddd"}]
matcher.add("usPhoneNum", [pattern])

doc= nlp("The phone number is 707-766-7600.")
matches = matcher(doc)
for mid, start, end in matches:
    print(doc[start:end])
707-766-7600
```

Voila! Our new pattern matched a phone number type entity as expected! Now, we'll deal with the cuisine type so that our chatbot can make a reservation. Let's see how to extract the cuisine type.

Extracting cuisine types

Extracting cuisine types is much easier than extracting a number of people or phone types; indeed, it's similar to extracting city entities. We can use a spaCy NER label directly for cuisine types – NORP. The NORP entity label refers to ethnic or political groups:

```
spacy.explain("NORP")
'Nationalities or religious or political groups'
```

Fortunately, cuisine names in our corpus coincide with nationalities. So, cuisine names are labeled as NORP by spaCy's NER.

First, let's have a look at some example utterances:

```
Is there a specific cuisine type you enjoy, such as Mexican,
Italian or something else?
I usually like eating the American type of food.
Find me Ethiopian cuisine in Berkeley.
I'm looking for a Filipino place to eat.
I would like some Italian food.
Malaysian sounds good right now.
```

Let's extract the entities of these utterances and examine how spaCy's NER labels cuisine types as follows:

```
for sent in sentences:
    doc = nlp(sent
    [(ent.text, ent.label_) for ent in doc.ents]
[('Mexican', 'NORP'), ('Italian', 'NORP')]
[('American', 'NORP')]
[('Ethiopian', 'NORP'), ('Berkeley', 'GPE')]
[('Filipino', 'NORP')]
[('Italian', 'NORP')]
[('Malaysian', 'NORP')]
```

Now, we are able to extract the city, date and time, number of people, and cuisine entities from user utterances. The result of the named entity extraction module we built here carries all the information the chatbot needs to provide to the reservation system. Here's an example utterance annotated with extracted entities:

```
I'd like to reserve an Italian place for 4 people by tomorrow
19:00 in Berkeley.
{
entities: {
    "cuisine": "Italian",
    "date": "tomorrow",
    "time": "19:00",
    "number_people": 4,
    "city": "Berkeley"
}
```

Here, we completed the first part of our semantic parsing, extracting entities. A full semantic parse needs an intent too. Now, we'll move on to the next section and do intent recognition with TensorFlow and Keras.

Intent recognition

Intent recognition (also called **intent classification**) is the task of classifying user utterances with predefined labels (intents). Intent classification is basically text classification. Intent classification is a well-known and common NLP task. GitHub and Kaggle host many intent classification datasets (please refer to the *References* section for the names of some example datasets).

In real-world chatbot applications, we first determine the domain our chatbot has to function in, such as finance and banking, healthcare, marketing, and so on. Then we perform the following loop of actions:

1. We determine a set of intents we want to support and prepare a labeled dataset of (utterance, label) pairs. We train our intent classifier on this dataset.

2. Next, we deploy our chatbot to the users and gather real user data.

3. Then we examine how our chatbot performed on real user data. At this stage, usually, we spot some new intents and some utterances our chatbot failed to recognize. We extend our set of intents with the new intents, add the unrecognized utterances to our training set, and retrain our intent classifier.

4. We go to *step 2* and perform *steps 2-3* until chatbot NLU quality reaches a good level of accuracy (> 0.95)

Our dataset is a real-world dataset; it contains typos and grammatical mistakes. While designing our intent classifiers – especially while doing pattern-based classification – we need to be robust to such mistakes.

We'll do the intent recognition in two steps: pattern-based text classification and statistical text classification. We saw how to do statistical text classification with TensorFlow and Keras in *Chapter 8, Text Classification with spaCy*. In this section, we'll work with Tensorflow and Keras again. Before that, we'll see how to design a pattern-based text classifier.

Pattern-based text classification

Pattern-based classification means classifying text by matching a predefined list of patterns to the text. We compare a precompiled list of patterns against the utterances and check whether there's a match.

An immediate example is **spam classification**. If an email contains one of the patterns, such as *you won a lottery* and *I'm a Nigerian prince*, then this email should be classified as spam. Pattern-based classifiers are combined with **statistical classifiers** to boost the overall system accuracy.

Contrary to statistical classifiers, pattern-based classifiers are easy to build. We don't need to put any effort into training a TensorFlow model at all. We will compile a list of patterns from our corpus and feed them to Matcher. Then, Matcher can look for pattern matches in utterances.

To build a pattern-based classifier, we first need to collect some patterns. In this section, we'll classify utterances with the NONE label. Let's see some utterance examples first:

```
No, Thanks
No, thank you very much.
That is all thank you so much.
No, that is all.
Nope, that'll be all. Thanks!
No, that's okay.
No thanks. That's all I needed help with.
No. This should be enough for now.
No, thanks.
No, thanks a lot.
No, thats all thanks.
```

By looking at these utterances, we see that the utterances with the NONE label follow some patterns:

- Most of the utterances start with No, or No..

- Patterns of saying *thank you* are also quite common. The patterns Thanks, thank you, and thanks a lot occur in most of the utterances in the preceding code.

- Some helper phrases such as that is all, that'll be all, that's OK, and this should be enough are also commonly used.

Based on this information, we can create three Matcher patterns as follows:

```
[{"LOWER": {"IN": ["no", "nope"]}}, {"TEXT": {"IN": [",",
"."]}}]
[{"TEXT": {"REGEX": "[Tt]hanks?"}}, {"LOWER": {"IN": ["you", "a
lot"]}, "OP": "*"}]
[{"LOWER": {"IN": ["that", "that's", "thats", "that'll",
```

```
"that11"]}}, {"LOWER": {"IN": ["is", "will"]}, "OP": "*"},
{"LOWER": "all"}]
```

Let's go over the patterns one by one:

- The first pattern matches token sequences no,, no., nope,, nope., No,, No., Nope,, and Nope.. The first item matches two tokens no and nope either in capitals or small letters. The second item matches the punctuation marks , and ..

- The second pattern matches thank, thank you, thanks, and thanks a lot, either in capitals or small letters. The first item matches thank and thanks s?. In regex syntax, the s character is optional. The second item corresponds to the words you and a lot, which can possibly follow thanks?. The second item is optional; hence, the pattern matches thanks and thank as well. We used the operator OP: * to make the second item optional; recall from *Chapter 4, Rule-Based Matching,* that Matcher supports operator syntax with different operators, such as * , +, and ?.

- The third pattern matches the token sequences that is all, that's all, thats all, and so on. Notice that the first item includes some misspelled words, such as thats and that11. We included the misspelled words on purpose, so the matching will be more robust to user typos.

Different combinations of the preceding three patterns will match the utterances of the NONE class. You can try the patterns by adding them to a Matcher object and see how they match.

Pro tip

While designing a rule-based system, always keep in mind that user data is not perfect. User data contains typos, grammatical mistakes, and wrong capitalization. Always keep robustness as a high priority and test your patterns on user data.

We made a statistical model-free classifier by making use of some common patterns and classified one intent successfully. How about the other two intents – FindRestaurants and ReserveRestaurant? Utterances of these two intents are semantically much more complicated, so we cannot cope with pattern lists. We need statistical models to recognize these two intents. Let's go ahead and train our statistical text classifiers with TensorFlow and Keras.

Classifying text with a character-level LSTM

In this section, we'll train a **character-level LSTM architecture** for recognizing the intents. We already practiced text classification with TensorFlow and Keras in *Chapter 8, Text Classification with spaCy*. Recall from this chapter that LSTMs are sequential models that process one input at one time step. We fed one word at each time step as follows:

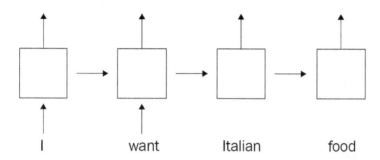

Figure 10.3 – Feeding one word to an LSTM at each time step

As we remarked in *Chapter 8, Text Classification with spaCy*, LSTMs have an internal state (you can think of it as a memory), so LSTMs can model the sequential dependencies in the input sequence by holding past information in their internal state.

In this section, we'll train a character-level LSTM. As the name suggests, we'll feed utterances character by character, not word by word. Each utterance will be represented as a sequence of characters. At each time step, we'll feed one character. This is what feeding the utterance from *Figure 10.3* looks like:

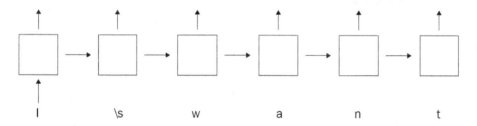

Figure 10.4 – Feeding the first two words of the utterance "I want Italian food"

We notice that the space character is fed as an input as well, because the space character is also a part of the utterance; for character-level tasks, there is no distinction between digits, spaces, and letters.

Let's start building the Keras model. We'll skip the data preparation stage here. You can find the complete code in the intent classification notebook https://github.com/PacktPublishing/Mastering-spaCy/blob/main/Chapter10/Intent-classifier-char-LSTM.ipynb.

We'll directly start with Keras' Tokenizer to create a vocabulary. Recall from *Chapter 8, Text Classification with spaCy*, that we use Tokenizer to do the following:

- Create a vocabulary from the dataset sentences.

- Assign a token ID to each token of the dataset.

- Transform input sentences to token IDs.

Let's see how to perform each step:

1. In *Chapter 8, Text Classification with spaCy*, we tokenized the sentences into words and assigned token IDs to words. This time, we'll break the input sentence into its characters, then assign token IDs to characters. Tokenizer provides a parameter named char_level. Here's the Tokenizer code for character-level tokenization:

   ```
   from tensorflow.keras.preprocessing.text import Tokenizer
   tokenizer = Tokenizer(char_level=True, lower=True)
   tokenizer.fit_on_texts(utterances)
   ```

2. The preceding code segment will create a vocabulary from the input characters. We used the lower=True parameter, so all characters of the input sentence are made lowercase by Tokenizer. After initializing the Tokenizer object on our vocabulary, we can now examine its vocabulary. Here are the first 10 items of the Tokenizer vocabulary:

   ```
   tokenizer.word_index
   {' ': 1, 'e': 2, 'a': 3, 't': 4, 'o': 5, 'n': 6, 'i': 7,
   'r': 8, 's': 9, 'h': 10}
   ```

   Just as with the word-level vocabulary, index 0 is reserved for a special token, which is the padding character. Recall from *Chapter 8, Text Classification with spaCy*, that Keras cannot process variable-length sequences; each sentence in the dataset should be of the same length. Hence, we pad all sentences to a maximum length by appending a padding character to the sentence end or sentence start.

3. Next, we'll convert each dataset sentence into token IDs. This is achieved by calling the `texts_to_sequences` method of Tokenizer:

```
utterances = tokenizer.texts_to_sequences(utterances)
utterances[0]
[17, 2, 9, 25, 1, 7, 1, 22, 3, 6, 4, 1, 7, 4, 1, 5, 6, 1,
4, 10, 2, 1, 28, 28, 4, 10]
```

4. Next, we'll pad all the input sentences to a length of `150`:

```
MAX_LEN = 150
utterances =\
  pad_sequences(utterances, MAX_LEN, padding="post")
```

We're ready to feed our transformed dataset into our LSTM model. Our model is a simple yet very efficient one: we placed a dense layer on top of a bidirectional LSTM layer. Here's the model architecture:

```
utt_in = Input(shape=(MAX_LEN,))
embedding_layer = Embedding(input_dim = len(tokenizer.
word_index)+1, output_dim = 100, input_length=MAX_LEN)
lstm =\
Bidirectional(LSTM(units=100, return_sequences=False))
utt_embedding = embedding_layer(utt_in)
utt_encoded = lstm(utt_embedding)
output = Dense(1, activation='sigmoid')(utt_encoded)
model = Model(utt_in, output)
```

A bidirectional LSTM layer means two LSTMs stacked on top of each other. The first LSTM goes through the input sequence from left to right (in a forward direction) and the second LSTM goes through the input sequence right to left (in a backward direction). For each time step, the outputs of the forward LSTM and backward LSTM are concatenated to generate a single output vector. The following figure exhibits our architecture with a bidirectional LSTM:

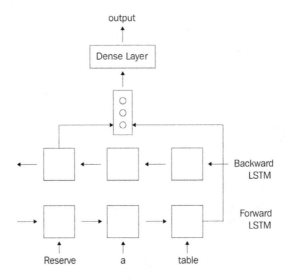

Figure 10.5 – Bidirectional LSTM architecture

5. Next, we compile our model and train it on our dataset by calling model.fit:

```
model.compile(loss = 'binary_crossentropy', optimizer =
"adam", metrics=["accuracy"])
model.fit(utterances, labels, validation_split=0.1,
epochs = 10, batch_size = 64)
```

Here, we compiled our model with the following:

a) Binary cross-entropy loss, because this is a binary classification task (we have two class labels).

b) The Adam optimizer, which will help the training procedure to run faster by arranging the size of the training steps. Please refer to the *References* section and *Chapter 8, Text Classification with spaCy*, for more information about the Adam optimizer.

c) Accuracy as our success metric. Accuracy is calculated by comparing how often the predicted label is equal to the actual label.

After fitting our model, our model gives a `0.8226` accuracy on the validation set, which is quite good.

Now, only one question remains: why did we prefer to train a character-level model this time? Character-level models definitely have some advantages:

- Character-level models are highly misspelling-tolerant. Consider the misspelled word *charactr* – whether or not the *e* is missing does not affect the overall sentence semantics that much. For our dataset, we will benefit from this robustness, as we have already seen spelling mistakes by users in our dataset exploration.

- The vocabulary size is smaller than a word-level model. The number of characters in the alphabet (for any given language) is fixed and low (a maximum of 50 characters, including uppercase and lowercase letters, digits, and some punctuation); but the number of words in a language is much greater. As a result, model sizes can differ. The main difference lies at the embedding layer; an embedding table is of size (`vocabulary_size, output_dim`) (refer to the model code). Given that the output dimensions are the same, 50 rows is really small compared to thousands of rows.

In this section, we were able to extract the user intent from the utterances. Intent recognition is the main step in understanding sentence semantics, but is there something more? In the next section, we'll dive into sentence-level and dialog-level semantics. More semantic parsing

This is a section solely on chatbot NLU. In this section, we'll explore sentence-level semantic and syntactic information to generate a deeper understanding of the input utterances, as well as providing clues to answer generation.

In the rest of this section, please think of the answer generation component as a black box. We provide the semantic parse of the sentence and it generates an answer based on this semantic parse. Let's start by dissecting sentence syntax and examining the subjects and objects of the utterances.

Differentiating subjects from objects

Recall from *Chapter 3, Linguistic Features*, that a sentence has two important grammatical components: a **subject** and an **object**. The subject is the person or thing that performs the action given by the verb of the sentence:

```
Mary picked up her brother.
He was a great performer.
It was rainy on Sunday.
```

```
Who is responsible for this mess?
The cat is very cute.
Seeing you makes me happy.
```

A subject can be a noun, a pronoun, or a noun phrase.

An object is the thing or person on which the subject performs the action given by the verb. An object can be a noun, a pronoun, or a noun phrase too. Here are some examples:

```
Lauren lost her book.
I gave her/direct object a book/indirect object.
```

So far, so good, but how does this information help us in our chatbot NLU?

Extracting the subject and the object helps us understand the sentence structure, hence adding one more layer to the semantic parse of the sentence. Sentence subject and object information directly relates to answer generation. Let's see some examples of utterances from our dataset:

```
Where is this restaurant?
```

The following figure shows the dependency parse of this utterance. The subject is the noun phrase this restaurant:

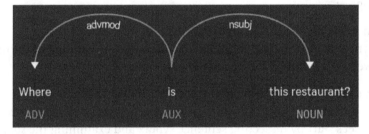

Figure 10.6 – Dependency parse of the example utterance

How can we generate an answer to this sentence? Obviously, the answer should have this restaurant (or the restaurant it refers to) as the subject. Some answers could be as follows:

```
The restaurant is located at the corner of 5th Avenue and 7th
Avenue.
The Bird is located at the corner of 5Th Avenue and 7Th Avenue.
```

What if the user puts `this restaurant` into the object role? Would the answer change? Let's take some example utterances from the dataset:

```
Do you know where this restaurant is?
Can you tell me where this restaurant is?
```

Obviously, the user is asking about the address of the restaurant again. The system needs to give the restaurant address information. However, this time, the subject of these sentences is you:

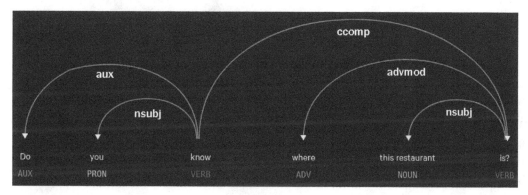

Figure 10.7 – Dependency parse of the first sentence

This question's subject is you, so the answer can start with an *I*. This a question sentence, hence the answer can start with a *yes/no* or the answer can just provide the restaurant's address directly. The following sentences are all possible answers:

```
I can give the address. Here it is: 5th Avenue, no:2
Yes, of course. Here's the address: 5th Avenue, no:2
Here's the address: 5Th Avenue, no:2
```

The same phrase, `the restaurant`, being the subject or the object doesn't affect the user's intent, but it affects the sentence structure of the answer. Let's look at the information more systematically. The semantic parses of the preceding example sentences look as follows:

```
{
utt: "Where is this restaurant?",
intent: "FindRestaurants",
entities: [],
structure: {
    subjects: ["this restaurant"]
```

```
    }
}
{
utt: "Do you know where is this restaurant is?",
intent: "FindRestaurants",
entities: [],
structure: {
    subjects: ["you"]
  }
}
```

When we feed these semantic parses to the answer generator module, this module can generate answers by taking the current utterance, the dialog history, the utterance intent, and the utterance's sentence structure (for the time being, only the sentence subject information) into account.

Here, we extracted the utterance's sentence structure information by looking at the utterance's dependency tree. Can a dependency parse provide us with more information about the utterance? The answer is yes. We'll see how to extract the sentence type in the next section.

Parsing the sentence type

In this section, we'll extract the sentence type of the user utterances. The grammar has four main sentence types, classified by their purpose:

```
Declarative: John saw Mary.
Interrogative: Can you go there?
Imperative: Go there immediately.
Exclamation: I'm excited too!
```

Sentence types in chatbot NLU are a bit different; we classify sentences according to the POS tag of the subject and objects as well as the purpose. Here are some sentence types that are used in chatbot NLU:

```
Question sentence
Imperative sentence
Wish sentence
```

Let's examine each sentence type and its structural properties. We start with question sentences.

Question sentences

A question sentence is used when the user wants to ask something. A question sentence can be formed in two ways, either by using an interrogative pronoun or by placing a modal/auxiliary verb at the beginning of the sentence:

```
How did you go there?
Is this the book that you recommended?
```

Hence, we also divide the question sentences into two classes, **wh-questions** and **yes/no questions**. As the name suggests, wh-questions start with a **wh-word** (a wh-word means an interrogative pronoun, such as where, what, who, and how) and yes/no questions are formed by using a modal/auxiliary verb.

How will this classification help us? Syntactically, yes/no questions should be answered with a yes or no. Hence, if our chatbot NLU passes a yes/no question to the answer generation module, the answer generator should evaluate this information and generate an answer that starts with a yes/no. Wh-questions aim to get information about the subject or objects, hence the answer generator module should provide information about the sentence subject or objects. Consider the following utterance:

```
Where is this restaurant?
```

This utterance generates the following dependency parse:

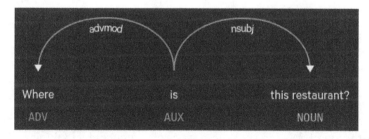

Figure 10.8 – Dependency parse of the example wh-question

Here, the subject of the utterance is this restaurant; hence the answer generator should generate an answer by relating Where and this restaurant. How about the following utterance:

```
Which city is this restaurant located?
```

The dependency parse of this utterance looks as follows:

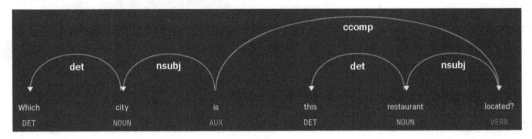

Figure 10.9 -- Dependency parse of the example wh-question

Here, the sentence structure is a bit different. `Which city` is the subject of the sentence and `this restaurant` is the subject of the clause. Here, the answer generation module should generate an answer by relating `which city` and `this restaurant`.

We will now move on to imperative sentence type.

Imperative sentence

Imperative sentences occur quite frequently in chatbot user utterances. An imperative sentence is formed by placing the main verb at the beginning of the sentence. Here are some utterance examples from our dataset:

```
Find me Ethiopian cuisine in Berkeley.
Find me a sushi place in Alameda.
Find a place in Vallejo with live music.
Please reserve me at 6:15 in the evening.
Reserve for six in the evening.
Reserve it for half past 1 in the afternoon.
```

As we see, imperative utterances occur quite a lot in user utterances, because they're succinct and to-the-point. We can spot these types of sentences by looking at the POS tags of the words: either the first word is a verb or the sentence starts with *please* and the second word is a verb. The following Matcher patterns match imperative utterances:

```
[{"POS": "VERB, "IS_SENT_START": True}]
[{"LOWER": "please", IS_SENT_START: True}, {"POS": "VERB"}]
```

How would the answer generator process these types of sentences? Imperative sentences usually include the syntactic and semantic elements to generate an answer; the main verb provides the action and is usually followed by a list of objects. Here's an example parse for the utterance `Find me Ethiopian cuisine in Berkeley`:

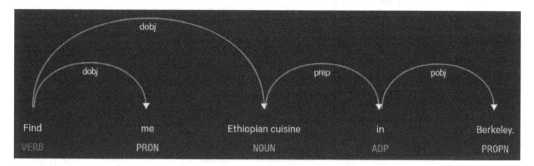

Figure 10.10 – Dependency parse of the example utterance

From the figure, we can see the syntactic components of this sentence as `Find` (the action), `Ethiopian cuisine` (an object), and `Berkeley` (an object). These components provide a clear template to the answer generator for generating an answer to this utterance: the answer generator should ask the restaurants database for matches of `Ethiopian cuisine` and `Berkeley` and list the matching restaurants.

Now we move on to the next sentence type, wish sentences. Let's see look at sentences in detail.

Wish sentences

Wish sentences are semantically similar to imperative sentences. The difference is syntactic: wish sentences start with phrases such as *I'd like to*, *Can I*, *Can you*, and *May I*, pointing to a wish. Here are some examples from our dataset:

```
I'd like to make a reservation.
I would like to find somewhere to eat, preferably Asian food.
I'd love some Izakaya type food.
Can you find me somewhere to eat in Dublin?
Can we make it three people at 5:15 pm?
Can I make a reservation for 6 pm?
```

Extracting the verb and the objects is similar to what we do for imperative sentences, hence the semantic parse is quite similar.

After extracting the sentence type, we can include it into our semantic parse result as follows:

```
{
utt: "Where is this restaurant?",
intent: "FindRestaurants",
entities: [],
structure: {
    sentence_type: "wh-question",
    subjects: ["this restaurant"]
  }
}
```

Now, we have a rich semantic and syntactic representation of the input utterance. In the next section, we'll go one step beyond the sentence-level semantics and go through the dialog-level semantics. Let's move on to the next section and see how we tackle dialog-level semantics.

Anaphora resolution

In this section, we'll explore the linguistic concepts of **anaphora** and **cohesion**. In linguistics, cohesion means the grammatical links that glue a text together semantically. This text can be a single sentence, a paragraph, or a dialog segment. Consider the following two sentences:

```
I didn't like this dress. Can I see another one
```

Here, the word one refers to the dress from the first sentence. A human can resolve this link easily. It's not so straightforward for software programs, though.

Also, consider the following dialog segment:

```
Where are you going?
To my grandma's.
```

The second sentence is completely understandable, though some parts of the sentence are missing:

```
I'm going to my grandma's house.
```

In written and spoken language, we use such **shortcuts** every day. However, resolving such shortcuts needs attention while programming, especially in chatbot NLU. Consider these utterances and dialog segments from our dataset:

Example 1:

```
- Do you want to make a reservation?
- Yes, I want to make one.
```

Example 2:

```
- I've found 2 Malaysian restaurants in Cupertino. Merlion
Restaurant & Bar is one.
- What is the other one?
```

Example 3:

```
- There's another restaurant in San Francisco that's called
Bourbon Steak Restaurant.
- Yes, I'm interested in that one.
```

Example 4:

```
- Found 3 results, Asian pearl Seafood Restaurant is the best
one in Fremont city, hope you like it.
- Yes, I like the same.
```

Example 5:

```
- Do you have a specific which you want the eating place to be
located at?
- I would like for it to be in San Jose.
```

Example 6:

```
- Would you like a reservation?
- Yes make it for March 10th.
```

All the highlighted parts of the preceding sentences and dialogs are examples of a linguistic event named **anaphora**. Anaphora means to look backward linguistically. An anaphora consists of two phrases: a phrase that refers to a phrase previously used in the context and the phrase that is referred to. Commonly used anaphoric words are one, more, same, it, and so on. Anaphora resolution means to resolve exactly the phrases anaphoric words point to.

How do we apply this information to our chatbot NLU then?

First of all, we need to determine whether an utterance involves an anaphora and whether we need an anaphora resolution. Consider the following dialog segment again:

```
Do you want to make a reservation?
Yes, I want to make one.
```

The dependency parse of the second utterance looks like this:

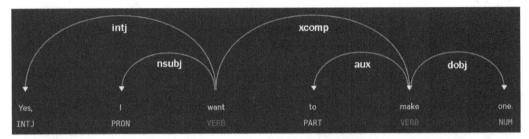

Figure 10.11 – Dependency parse of the example utterance

First of all, one appears as the direct object of the sentence and there are no other direct objects. This means that one should be an anaphora. In order to resolve what one refers to, we'll look back to the first utterance of the dialog. The following dependency parse belongs to the first utterance, Do you want to make a reservation?:

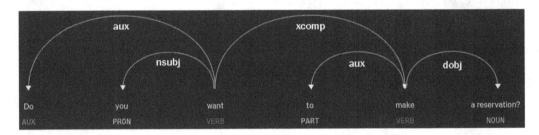

Figure 10.12 – Dependency parse of the example utterance

If we look at *Figure 10.11*, we see that the sentence has a direct object, a reservation, so one should refer to a reservation. Then, we can arrange the resulting semantic parse as follows:

```
{
utt: "Where is this restaurant?",
intent: "ReserveRestaurant",
entities: [],
structure: {
```

```
    sentence_type: "declarative",
    subjects: ["one"]
    anaphoras: {
        "one": "a reservation"
        }
    }
}
```

Replacing one with a reservation makes the sentence intent clearer. In our chatbot NLU, we only have two intents, but what if there are more intents, such as reservation cancellation, refunds, and so on? Then I want to make one can mean making a cancellation or getting a refund as well.

Therefore, we make anaphora resolution come before the intent recognition and feed the full sentence, where anaphora words are replaced with the phrases they refer to. This way, the intent classifier is fed with a sentence where the direct object is a noun phrase, not one of the words one, same, it, or more, which do not carry any meaning on their own.

Now after extracting meaning (by extracting intent) statistically with Keras, in this section you learned ways of processing sentence syntax and semantics with special NLU techniques. You're ready to combine all the techniques you know and design your own chatbot NLU for your future career. This book started with linguistic concepts, continued with statistical applications, and in this chapter, we combined it all. You're ready to keep going. In all the NLU pipelines you'll design, always try to look at the problem from a different view and remember what you learned in this book.

Summary

That's it! You made it to the end of this exhaustive chapter and also to the end of this book!

In this chapter, we designed an end-to-end chatbot NLU pipeline. As a first task, we explored our dataset. By doing this, we collected linguistic information about the utterances and understood the slot types and their corresponding values. Then, we performed a significant task of chatbot NLU, entity extraction. We extracted several types of entities such as city, date/time, and cuisine with the spaCy NER model as well as Matcher. Then, we performed another traditional chatbot NLU pipeline task – intent recognition. We trained a character-level LSTM model with TensorFlow and Keras.

In the last section, we dived into sentence-level and dialog-level semantics. We worked on sentence syntax by differentiating subjects from objects, then learned about sentence types and finally learned about the linguistic concept of anaphora resolution. We applied what we learned in the previous chapters, both linguistically and statistically, by combining several spaCy pipeline components such as NER, dependency parsers, and POS taggers.

References

Here are some references for this chapter:

On voice assistant products:

- Alexa developer blog: `https://developer.amazon.com/blogs/home/tag/Alexa`

- Alexa science blog: `https://www.amazon.science/tag/alexa`

- Microsoft's publication on chatbots: `https://academic.microsoft.com/search?q=chatbot`

- Google Assistant: `https://assistant.google.com/`

Keras layers and optimizers:

- Keras layers: `https://keras.io/api/layers/`

- Keras optimizers: `https://keras.io/api/optimizers/`

- An overview of optimizers: `https://ruder.io/optimizing-gradient-descent/`

- Adam optimizer: `https://arxiv.org/abs/1412.6980`

Datasets for conversational AI:

- Taskmaster from Google Research: `https://github.com/google-research-datasets/Taskmaster/tree/master/TM-1-2019`

- Simulated Dialogue dataset from Google Research: `https://github.com/google-research-datasets/simulated-dialogue`

- Dialog Challenge dataset from Microsoft: `https://github.com/xiul-msr/e2e_dialog_challenge`

- Dialog State Tracking Challenge dataset: `https://github.com/matthen/dstc`

Packt.com

Subscribe to our online digital library for full access to over 7,000 books and videos, as well as industry leading tools to help you plan your personal development and advance your career. For more information, please visit our website.

Why subscribe?

- Spend less time learning and more time coding with practical eBooks and Videos from over 4,000 industry professionals

- Improve your learning with Skill Plans built especially for you

- Get a free eBook or video every month

- Fully searchable for easy access to vital information

- Copy and paste, print, and bookmark content

Did you know that Packt offers eBook versions of every book published, with PDF and ePub files available? You can upgrade to the eBook version at packt.com and as a print book customer, you are entitled to a discount on the eBook copy. Get in touch with us at customercare@packtpub.com for more details.

At www.packt.com, you can also read a collection of free technical articles, sign up for a range of free newsletters, and receive exclusive discounts and offers on Packt books and eBooks.

Other Books You May Enjoy

If you enjoyed this book, you may be interested in these other books by Packt:

Python Natural Language Processing Cookbook

Zhenya Antić

ISBN: 978-1-83898-731-2

- Become well-versed with basic and advanced NLP techniques in Python
- Represent grammatical information in text using spaCy, and semantic information using bag-of-words, TF-IDF, and word embeddings
- Perform text classification using different methods, including SVMs and LSTMs
- Explore different techniques for topic modeling such as K-means, LDA, NMF, and BERT
- Work with visualization techniques such as NER and word clouds for different NLP tools
- Build a basic chatbot using NLTK and Rasa
- Extract information from text using regular expression techniques and statistical and deep learning tools

Getting Started with Google BERT

Sudharsan Ravichandiran

ISBN: 978-1-83882-159-3

- Understand the transformer model from the ground up

- Find out how BERT works and pre-train it using masked language model (MLM) and next sentence prediction (NSP) tasks

- Get hands-on with BERT by learning to generate contextual word and sentence embeddings

- Fine-tune BERT for downstream tasks

- Get to grips with ALBERT, RoBERTa, ELECTRA, and SpanBERT models

- Get the hang of the BERT models based on knowledge distillation

- Understand cross-lingual models such as XLM and XLM-R

- Explore Sentence-BERT, VideoBERT, and BART

Packt is searching for authors like you

If you're interested in becoming an author for Packt, please visit authors. packtpub.com and apply today. We have worked with thousands of developers and tech professionals, just like you, to help them share their insight with the global tech community. You can make a general application, apply for a specific hot topic that we are recruiting an author for, or submit your own idea.

Leave a review - let other readers know what you think

Please share your thoughts on this book with others by leaving a review on the site that you bought it from. If you purchased the book from Amazon, please leave us an honest review on this book's Amazon page. This is vital so that other potential readers can see and use your unbiased opinion to make purchasing decisions, we can understand what our customers think about our products, and our authors can see your feedback on the title that they have worked with Packt to create. It will only take a few minutes of your time, but is valuable to other potential customers, our authors, and Packt. Thank you!

Index

www.ingramcontent.com/pod-product-compliance
Lightning Source LLC
Chambersburg PA
CBHW062054050326
40690CB00016B/3087